Security Electronics Circuits Manual

R. M. MARSTON

Newnes

OXFORD BOSTON JOHANNESBURG MELBOURNE NEW DELHI SINGAPORE

Newnes
An imprint of Butterworth-Heinemann
Linacre House, Jordan Hill, Oxford OX2 8DP
225 Wildwood Avenue, Woburn, MA 01801-2041
A division of Reed Educational and Professional Publishing Ltd

A member of the Reed Elsevier plc group

First published 1998

British Library Cataloguing in Publication Data
A catalogue record for this book is available from the British Library

Library of Congress Cataloguing in Publication Data
A catalogue record for this book is available from the Library of Congress

ISBN 0 7506 3007 8

Composition by Scribe Design, Gillingham, Kent
Printed and bound in Great Britain by Biddles Ltd, Guildford and Kings Lynn

Contents

Contents

Preface

Modern electronic security circuits and systems range in complexity from the simple electronic door-bell to the ultra-sophisticated wireless burglar alarm system that comes complete with an array of passive infra-red (PIR) movement detectors and contact sensors plus full remote-control and sensor-monitoring facilities. Electronic security systems have a wide range of applications in the home, in industry and commerce, and in automobiles. They can be designed to be activated by physical contact or by body proximity, or by variations in heat, light, or infra-red radiation levels, or in voltage, current, resistance or some other electrical property. This new wide-ranging 55 000-word manual explains – with the aid of 231 illustrations – the operating principles of the most important types of modern electronic security systems and – where applicable – provides the reader with a wide range of practical application circuits.

The manual is split into eight chapters. The opening chapter gives a concise description of electronic security system basic principles and devices, and all subsequent chapters describe practical security systems and circuits. Chapter 2 deals with 'contact-operated' circuits, which activate when an electrical switch or wiring circuit is opened or closed. Chapter 3 deals with optoelectronic circuits that are activated by visible light or by infra-red radiation, and Chapter 4 describes modern anti-burglary circuits and systems.

Chapter 5 of the book looks at temperature-sensitive security circuits, and Chapters 6 and 7 look at circuits that are meant for use in instrumentation applications, or are specifically designed for use in automobiles. The final chapter looks at a miscellaneous collection of practical security circuits that are activated by the presence of a liquid, steam, or gas, by sound, by the failure of AC power supplies, by the close or near proximity of a person or object, by human touch, or by the breaking of an ultrasonic beam, etc.

This book, like all others in the *Newnes Circuits Manual Series*, is aimed at practical design engineers, technicians and experimenters, but will also be of great interest to all amateurs and students of electronics. It deals with its subject in an easy-to-read, down-to-earth, mainly non-mathematical but very comprehensive manner. Each chapter explains the basic principles of its subject and – where appropriate – presents the reader with a wide selection of practical application circuits, all of which have been designed and fully evaluated by the author.

Throughout the volume, great emphasis is placed on practical 'user' information and circuitry; most of the ICs and other devices used in the practical circuits are modestly priced and readily available types, with universally recognized type numbers. All of the book's diagrams have been generated by the author, using a basic 'Corel DRAW 3' graphics package.

In this book, the values of resistors and capacitors, etc., are notated in the International style that is now used throughout most of the western world, but which may not be familiar to some 'hobbyist' readers in the USA. Such readers should thus note the following points regarding the use of the International notation style *in circuit diagrams:*

(1) In resistance notation, the symbol R represents *units* of resistance, k represents *thousands* of units, and M represents *millions* of units. Thus, 10R = 10Ω, 47k = 47kΩ, 47M = 47MΩ.
(2) In capacitance notation, the symbols μ, n (= 1000pf), and p are used as basic multiplier units. Thus, 47μ = 47μF, 10n = 0.01μF, and 47p = 47pF.
(3) In the international notation system, decimal points are not used in notations and are replaced by the multiplier symbol (such as V, k, n, μ, etc.) applicable to the individual component value. Thus, 4V7 = 4.7V, 4k7 = 4.7kΩ, 4n7 = 4.7nF, and 1n0 = 1.0n.

R. M. Marston, 1998

1

Security system basics

Any system that provides its owner/user with a reasonable degree of protection against one or more real or imagined dangers, threats, or nuisances (such as physical attack, theft of property, unwanted human or animal intrusion, machine breakdown, or risks from fire, electric shock, or vermin infestation, etc.) can be described as a 'security' system. An 'electronic' security system is one in which the system's actions are heavily dependent on electronic circuitry; simple examples of such systems are electronic door-bells, key-pad door locks, and domestic burglar alarms. The opening chapter of this book starts off by explaining electronic security system basic principles and then goes on to describe a wide variety of devices that can be used within modern electronic security systems. Later chapters show practical examples of various specific types of low- to medium-complexity electronic security systems and circuits.

Electronic security system basics

All electronic security systems consist of the basic elements shown in *Figure 1.1*. Here, one or more 'danger' sensing units are placed at the front of the system and generate some kind of electrical output when danger is sensed. The output of the sensor unit is fed, via a data link, to a decision-making signal processing unit, and this unit's output is fed, via another data link, to a 'danger' response unit such as an alarm or an electromechanical trigger or shutdown device. Note in *Figure 1.1* that each of the system's three major elements is shown using its own power supply, but that in practice two or more elements may share a single power supply.

Figures 1.2 to *1.5* show, in basic form, four different low- to medium-complexity types of security system. The first of these (*Figure 1.2*) is a simple electronic door-bell or shop-entry alarm system, in which the 'danger' sensor

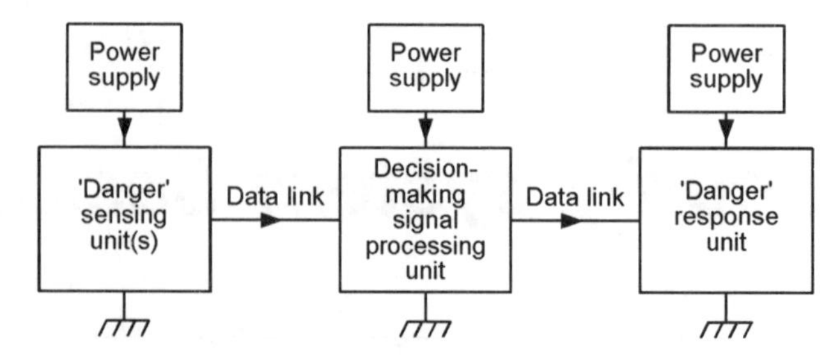

Figure 1.1 *Basic elements of an electronic security system*

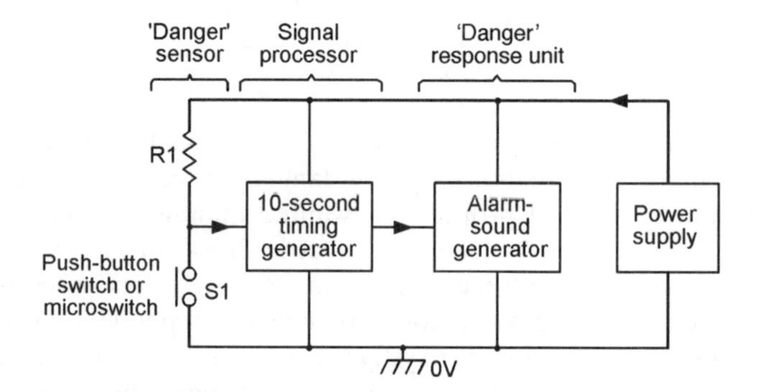

Figure 1.2 *Electronic door-bell or shop-entry system*

is a push-button switch in the case of the door-bell system or a door-mounted microswitch (or a pressure mat switch, etc.) in the case of the shop-entry system. In both cases, the circuit action is such that when switch S1 closes it activates a timing generator that turns on an alarm sound generator for a period of 10 seconds, irrespective of the actual duration of the switch closure, and repeats this action each time that S1 is closed. Ideally, this type of circuit draws zero quiescent current. Note in the case of the door-bell circuit that the 'danger' sensor (S1) is operated voluntarily by the unknown visitor, in a deliberate effort to attract the attention of the householder, but that in the case of the shop-entry circuit S1 is operated involuntarily by the visitor, and warns the shopkeeper of the presence of a potential customer or thief.

Figure 1.3 shows a simple domestic burglar alarm circuit. Here, the main alarm system is enabled by closing key-operated switch S2, and the S1 'danger' sensor actually consists of any desired number of series-connected normally-closed switches (usually reed-and-magnet types) that are each wired to a protected door or window, so that the composite S1 switch opens when any protected door or window is opened or a break occurs in S1's wiring. Under this condition, R1 pulls the input of the transient-suppressing

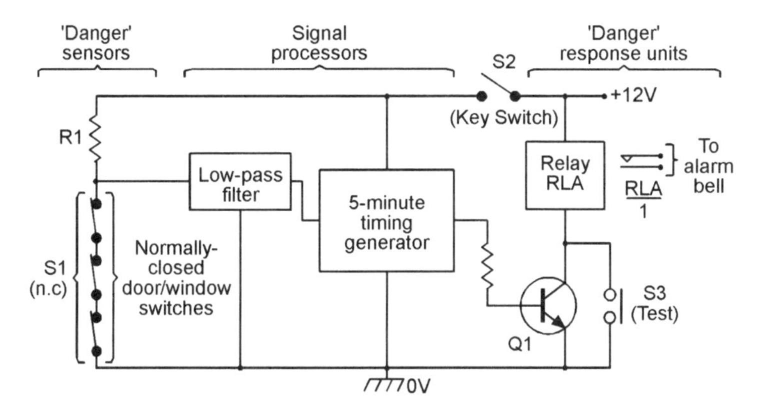

Figure 1.3 *Simple domestic burglar alarm system*

low-pass filter high, and after a brief delay (usually about 200mS) the filter output triggers the 5-minutes timer generator, which turns on relay RLA via transistor Q1 and thereby activates an external alarm bell or siren via the relay's RLA/1 contacts. Once activated, the relay and alarm turn off automatically at the end of the 5-minute timing period, but can be turned off or reset at any time by opening key-switch S2. The alarm can be tested at any time, with or without closing S2, via push-button switch S3, which closes RLA directly.

Figure 1.4 shows, in pictorial form, a modern passive infra-red (PIR) movement detector system that can be used to automatically sound an alarm or turn on floodlights when a person enters the PIR detection field (the PIR has a typical maximum range of 12 metres and the field has a vertical span of about 15 degrees and a horizontal span of 90 to 180 degrees). The PIR unit detects the small amounts of infra-red radiation generated by human body heat, but gives an 'alarm' output only when the heat source *moves*

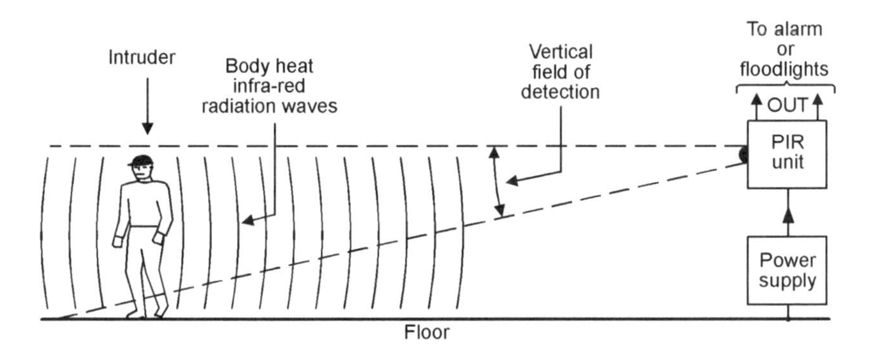

Figure 1.4 *Passive infra-red (PIR) movement detector system*

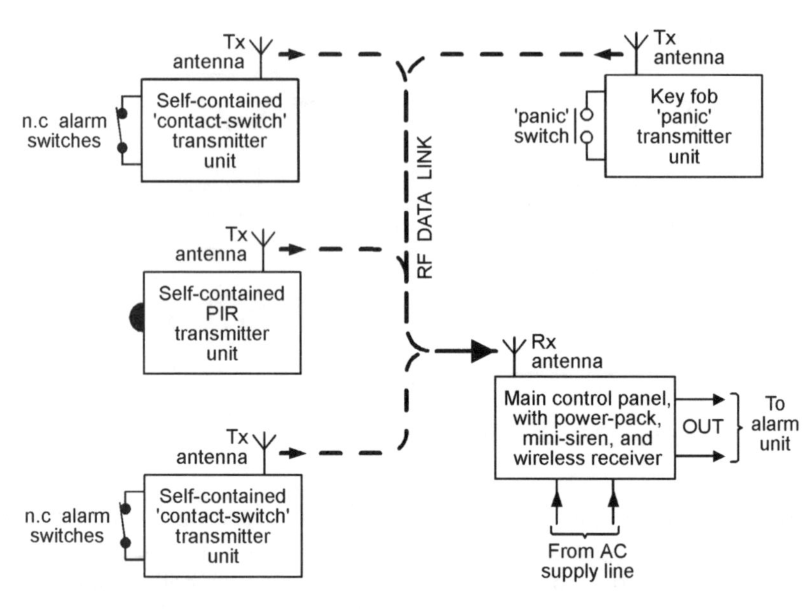

Figure 1.5 *Wireless burglar alarm system*

significantly within the detection field. Most PIR units have good immunity to false alarms; some types incorporate an output relay that is normally closed (turned on) but opens (turns off) when an intruder is detected or the unit's power supply fails or is removed; units of this latter type typically need a 12V DC supply and consume a quiescent current of about 20mA. PIR units are widely used to give room or area protection in modern burglar alarm systems.

Figure 1.5 shows, in simplified form, the basic elements of a modern domestic 'wireless' burglar alarm system, in which the data link between the various major parts of the system takes the form of a coded RF (usually 418MHz or 458MHz) signal, thus greatly easing installation problems. The heart of the system is the main control panel, which houses a wireless receiver and decoder and control logic, plus a high-power mini-siren, and has an output that can activate an external high-power siren and light-strobe alarm unit. The system's 'danger' sensing units each house a small RF transmitter and antenna that sends out a coded signal under a danger condition; each of the units are designed to give a minimum of six months of normal operation from a small battery.

Most domestic wireless burglar alarm systems can be used to monitor a maximum of four to six zones (individual protected areas) via suitable sensing units. The sensing units come in three basic types; 'contact-switch' types transmit a danger signal when one or more series-connected normally-closed switches are opened, and can be used to protect a zone of any desired size; 'PIR' types transmit a danger signal when a human moves within the visual field of the PIR unit, and can be used to protect a zone of limited size; 'panic' types transmit a danger signal when a key-fob button is pressed, and can be used to protect a person against sudden physical attack or threat whenever they are within communication range of the system's receiver (control panel) unit. All three types of sensing unit also send out monitoring signals that give warnings of failing battery power or deliberate interference, etc., and the wireless burglar alarm system thus offers a high degree of security.

Note that simple electronic security systems such as those shown in *Figures 1.2* and *1.3* can be easily and cheaply built on a DIY basis, but that it is not cost-effective to build a PIR unit of the *Figure 1.4* type as a DIY project, or cost-effective or legal (because the RF transmitters must be certified by an approved state or national body) to build (rather than buy) a *Figure 1.5* type of wireless burglar alarm system as a pure DIY project. Commercial PIR units and wireless burglar alarm units can, however, easily be used as special elements that can be incorporated in a wide variety of DIY security *systems*.

Security system reliability

The most important parameter of any practical electronic security system is its *reliability* in performing its designated task. Specifically, all such systems must be easy to use, difficult to disable, and have good immunity against malfunctioning and the generation of false alarms (which very quickly destroy the user's confidence in the system). The degree and types of reliability required from a security system vary with the level of security that the system is designed to provide. Domestic burglar alarm systems (in which only a few family members have access to the major functional parts of the system) have, for example, relatively low anti-tamper requirements, but anti-burglary systems used in large shops and stores – in which the public have easy access to many protected areas during normal 'opening' hours – have very high levels of anti-tamper requirement.

The overall reliability of any electronic security system is greatly influenced by the nature of its major system elements, i.e. by its danger sensing units and its data links, etc. Simple electromechanical danger sensors such as reed switches and pressure pad switches have, for example, far greater intrinsic levels of reliability than electronic sensors such as ultrasonic, microwave, and simple light-beam intrusion detectors, but electronic key-pad security switches usually have far greater reliability than the mechanical key-switches that they are designed to replace, and so on. To gain a useful insight into this subject, the reader needs a good understanding of the wide variety of elements that are used in modern electronic security systems.

Security system elements

All electronic security systems consist – as shown in *Figure 1.1* – of one or more 'danger' sensing units that generate some kind of electrical output when danger is sensed, and which feed that output – via a data link and a decision-making signal processing unit – to a 'danger' response unit such as an alarm or an electromechanical trigger or shutdown device. Apart from the actual signal processing unit, the three other major elements of any electronic security system are thus the sensing unit, the data link, and the response unit, and each of these elements may take an electromechanical, electrical, or an electronic form. Each of these three basic elements are available in a variety of guises, and the most important of these are described in the remaining sections of this chapter.

Electromechanical sensors

Simple switches

The simplest and most widely used electromechanical sensors are ordinary electrical switches of the various types shown in *Figures 1.6(a)* to *1.6(e)*. The types shown in *(a)* to *(d)* are linear pressure-operated types and may take normal manually-operated forms or may be microswitches that are activated by the mechanical movement of a door, window or machine part, etc. The *(e)* type is a rotary multi-step pressure-operated switch that is (normally)

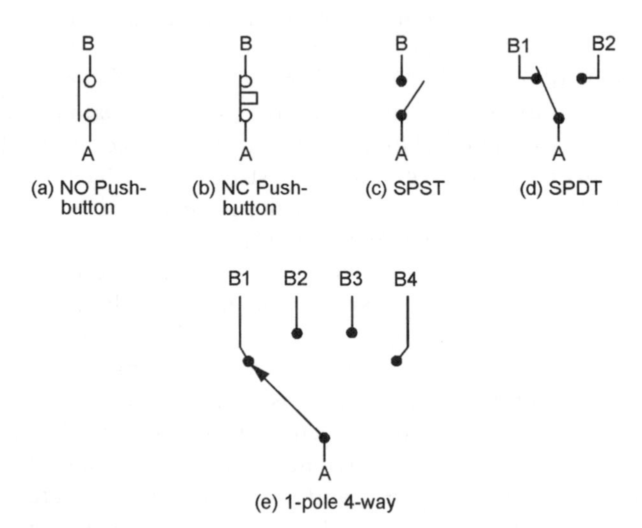

Figure 1.6 *Five basic switch configurations*

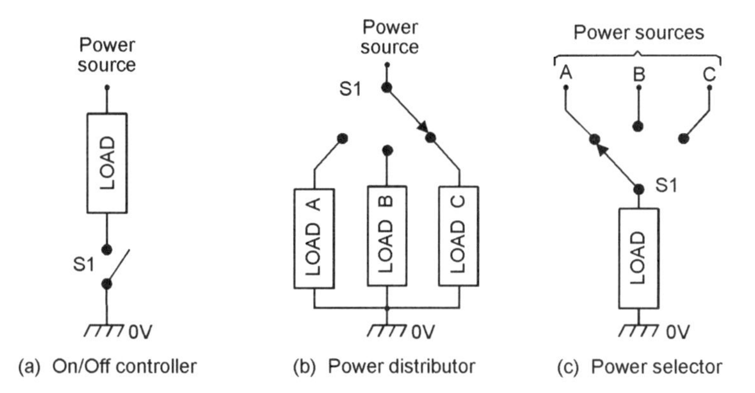

Figure 1.7 *Three basic types of power (or signal) switching circuit*

activated manually. The sensor shown in *(a)* is a normally-open (NO or n.o.) push-button switch, *(b)* is a normally-closed (NC or n.c.) push-button switch, *(c)* is a single-throw single-pole (SPST) toggle switch, *(d)* is a single-pole double-throw (SPDT) or 'change-over' toggle switch, and *(e)* is a single-pole 4-way rotary switch.

Figure 1.7 shows three basic ways of using normal electrical switches in power (or signal) switching applications. In *(a)*, a SPST switch is used as an on/off controller to switch power to a single load, in *(b)* a 1-pole 3-way switch is used as a power distributor to switch power to any one of three loads, and in *(c)* is used as a power selector, to connect any one of three power sources to a single load.

Switched-output electromechanical sensors are available in a variety of basic types, including temperature-sensitive thermostats, orientation-sensitive 'tilt' and 'tip-over' switches, pressure-sensitive 'mat' switches, key-operated security switches, and time-sensitive 'timer' switches, all of which are shown in basic form in *Figures 1.8* to *1.10*.

Thermostats

Thermostats are temperature-activated on/off switches that usually work on the 'bimetal' principle illustrated in *Figure 1.8(a)*, in which the bimetal strip consists of two bonded layers of conductive metal with different coefficients of thermal expansion, thus causing the strip to bend in proportion to temperature and to make (or break) physical and electrical contact with a fixed switch contact at a specific temperature. In practice, the bimetal element may be in strip, coiled, or snap-action conical disc form, depending on the application, and the thermal 'trip' point may or may not be adjustable. *Figures 1.8(b)* and *(c)* show the symbols used to represent fixed and variable thermostats. A variety of thermostats are readily available, and can easily be

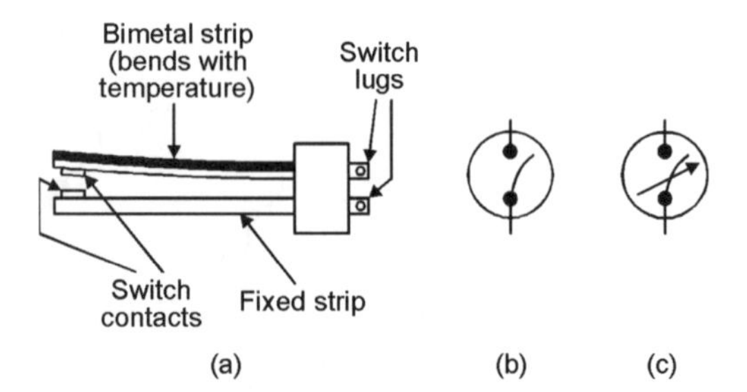

Figure 1.8 *Basic construction of a simple bimetal thermostat (a), and symbols for (b) fixed and (c) variable thermostats*

used in automatic temperature control or danger-warning (fire or frost) applications. Their main disadvantage is that they suffer from hysteresis; typically, a good quality adjusted thermostat may close when the temperature rises to (say) 21°C but not re-open again until it falls to 19.5°C.

Tilt switches

Figure 1.9(a) illustrates the basic construction and operating principle of a mercury tilt switch, which (in this example) consists of a cigar-shaped cavity that is formed within a block made of two electrically-connected metal end contacts and a central metal contact, which are separated by insulating sections. The cavity holds a mercury globule, which rests on the central contact but is insulated from the end contacts when the switch is horizontal, but rolls and touches one or other of the end contacts (and also the central contact) if the switch is tilted significantly (typically by more that 10 degrees) out of the horizontal. The mercury 'switch' is thus normally open, but closes when tilted, and can be used to activate an alarm if an attempt is made to move a normally-stationary protected item such as a TV, PC, or hi-fi unit, etc.

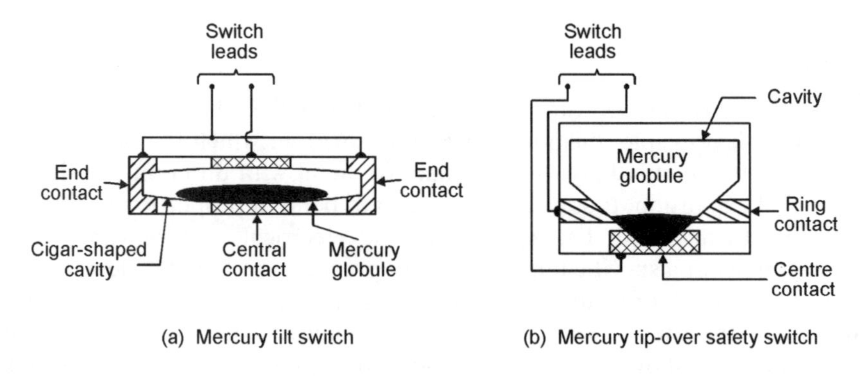

Figure 1.9 *Basic construction of mercury tilt (a) and tip-over (b) switches*

Tip-over switches

Figure 1.9(b) illustrates the basic construction and operating principle of a mercury tip-over safety switch. In this case the cavity is fairly steep-sided, and the construction is such that the mercury globule touches both a ring contact and a centre contact when the unit is vertical, and thus acts as a closed switch, but breaks this contact and acts as an open switch when the unit is tilted heavily (typically by more than 40 degrees) out of the vertical position. One common application of this type of switch is in free-standing electric heaters, where the switch is built into the unit and wired in series with its power lead, so that the appliance automatically turns off if it is accidentally knocked over.

Pressure mat switches

Figures 1.10(a) and *1.10(b)* illustrate the general appearance and basic construction of a pressure mat switch, which is designed to be hidden under a mat or carpet and acts as a normally-open switch that closes if a person steps heavily on any part of the switch. The device consists of two sheets of metal foil that are normally held apart by a perforated sheet of foam plastic; this sandwich is encased in a hermetically sealed plastic envelope; when a person treads on the envelope their weight compresses the foam plastic, and the metal foils make electrical contact via the foam sheet's perforations. Pressure mat switches are widely used in domestic and commercial burglar

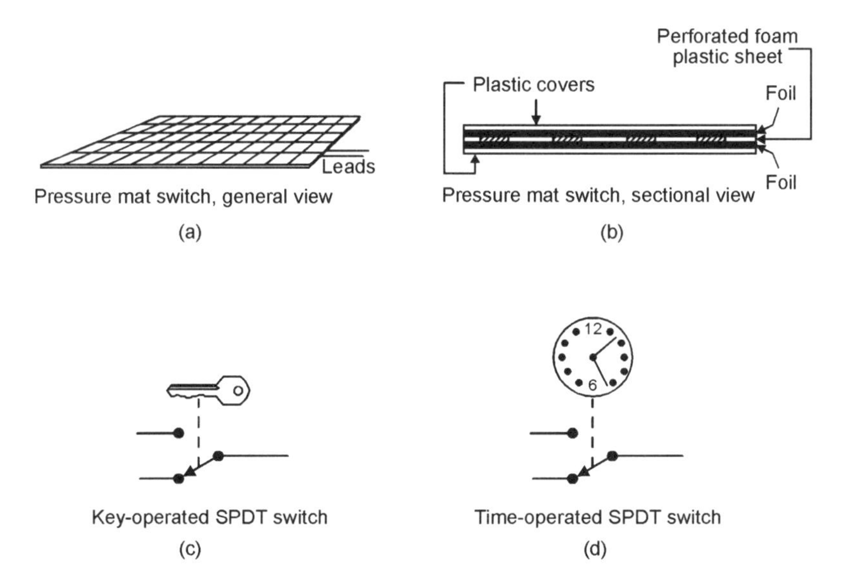

Figure 1.10 *General (a) and sectional (b) views of a pressure mat switch, and symbolic representations of (c) key-operated and (d) time-operated SPDT switches*

alarm systems; most such switches have four output wires; the two 'switch' wires have partly-bared ends; the other two wires are not bared, are internally shorted together, and serve a n.c anti-tamper function in which an alarm system activates if the sensor wiring is cut (see the *Data links* section of this chapter), and can be ignored in most domestic applications.

Key switches

Figure 1.10(c) shows a symbolic representation of a simple key-operated SPST electric switch, in which the switch arm is moved by turning a Yale-type key in a matching tumbler mechanism. Switches of this basic type are available in many different switch and key-type styles, and are widely used in security applications in buildings and vehicles, and on items such as PCs and burglar alarm control units. The most important parameter of a key-switch (or of any type of key-operated lock) is its number of 'differs' or possible key profiles; Yale-type switches have a number of pins (usually five) which must each be raised to a certain level by the key to allow the switch to operate; usually, each pin has three possible levels, and a simple 5-pin key switch thus has 243 (= 3^5) differs; if the key's shaft also carries two long grooves that must match the lock's face plate and offer (say) a further 9 differs, the total number of differs is raised to 2187.

Time switches

Figure 1.10(d) shows a symbolic representation of a simple analogue time-operated SPST electric switch, in which the switch arm is moved by a mechanical (clockwork or slow-release), electrical (current-heated thermostat) or electromechanical (synchronous motor plus gearbox) timing mechanism. Switches of this basic type are available in many different switch styles, with many different timing ranges, and are widely used in light-switching and solenoid-operating security applications.

Reed switches

One of the most useful types of switched-output electromechanical sensor devices is the 'reed' switch, which activates in the presence of a suitable magnetic field and is particularly useful in proximity-detector applications. *Figure 1.11* shows the basic structure of a reed switch, which consists of a springy pair of opposite-polarity magnetic reeds with plated low-resistance contacts, sealed into a glass tube filled with protective gasses. The opposing magnetic fields of the reeds normally hold their contacts apart, so they act as an open switch, but these fields can by nulled or reversed by placing the reeds within an externally-generated magnetic field (see *Figure 1.12*), so that the reed then acts as a closed switch.

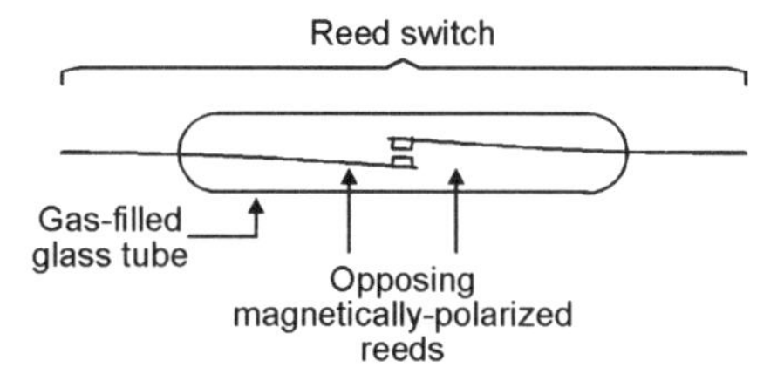

Figure 1.11 *Basic structure of a reed switch*

A reed switch can be activated by placing its reeds within an externally-generated magnetic field, which can be derived from either an electric coil that surrounds the glass tube, as in the 'reed relay' diagram of *Figure 1.12(a)*, or by a permanent magnet placed within a few millimetres of the tube, as shown in *Figure 1.12(b)*. Reed relays are used in the same way as normal relays, but typically have a drive-current sensitivity ten times better than a standard relay. Reed-and-magnet combinations are very useful in proximity-detector applications in security and safety systems, etc., as illustrated in *Figure 1.13*.

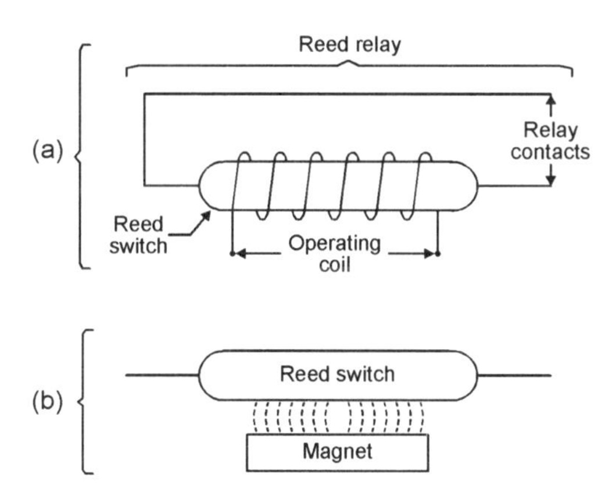

Figure 1.12 *Reed switch operated by (a) coil or (b) magnet*

Figure 1.13 shows a method of using a reed and magnet to give burglar protection to a door or window. Here, the reed switch is embedded in a door or window frame, and the activating magnet is embedded adjacent to it in the actual door or window so that the reed switch changes state whenever the door/window is opened or closed. The reed switch can thus be used to activate an alarm circuit whenever a protected door/window is opened. In practice, the reed and magnet may take the basic forms shown in *Figure 1.12(b)*, or may be encapsulated in special housings that can easily be screwed to – or embedded in – the frame/body of the door/window.

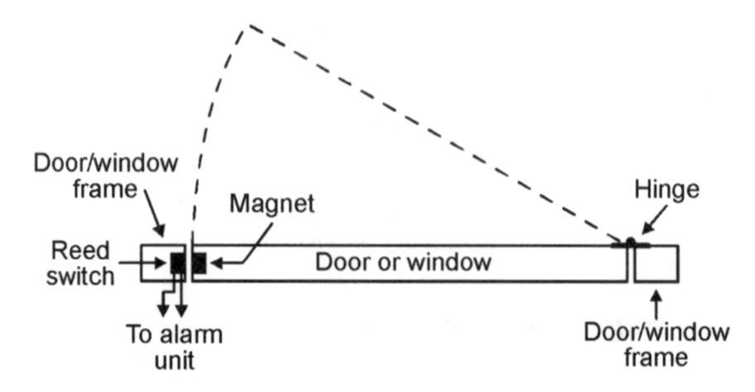

Figure 1.13 *Method of using a reed switch/magnet combination to give burglar protection to a door or window*

Basic alarm switching circuits

Several switched-output sensor devices can be used to activate an alarm bell or other device by connecting them in one or other of the basic modes shown in *Figure 1.14*. In *(a)*, the switches are wired in series, so that the alarm sounds only when all three switches are closed at the same moment. In *(b)*, the switches are wired in parallel, and the alarm sounds whenever any one of the switches is closed. In most practical alarm systems, a combination of series and parallel switching is used, as shown in the example of *Figure 1.15*. In this case, the alarm system is enabled (made alert) by closing series-connected time switch S1 and key switch S2; once enabled, the alarm bell can be activated by closing any of the parallel-connected S3 to S5 switches. In burglar alarm systems, important intrusion-sensing switches should be n.c.

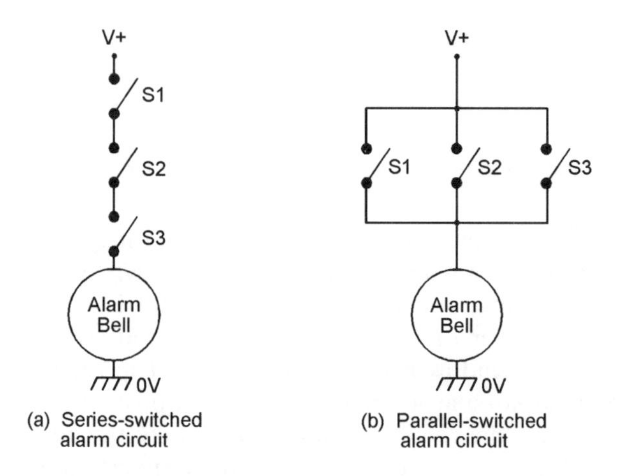

Figure 1.14 *An alarm bell can be activated by several switches wired (a) in series or (b) in parallel*

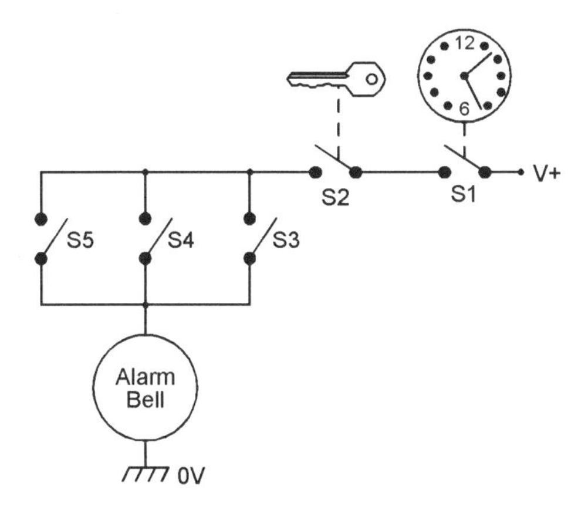

Figure 1.15 *Simple security alarm, using a combination of series- and parallel-connected switches*

types that are wired in series and used in the basic manner already shown in *Figure 1.3*, so that the alarm activates if any switch opens or if its wires are cut; R1 should have a high value (typically several megohms), to give low quiescent current consumption.

Electrical sensor devices

Thermistors

A thermistor is a passive resistor device with a resistance value that is highly sensitive to the device's temperature. Practical thermistors are available in rod, disc, and bead forms, and with either positive or negative temperature coefficients (known as PTC and NTC types respectively). Unlike electro-mechanical thermostats, they do not suffer from hysteresis problems, and are thus suitable for use in a variety of precision temperature sensing and switching applications. *Figure 1.16* shows two alternative symbols that can be used to represent a thermistor. In most practical applications, thermistors are used in conjunction with electronic circuitry that gives a switch-type output when the thermistor temperature goes above (or below) a pre-set limit. Thermistors have typical operating temperature ranges of –40°C to +125°C.

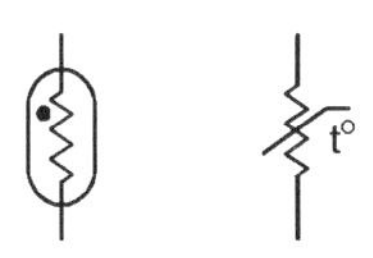

Figure 1.16 *Symbols commonly used to represent a thermistor*

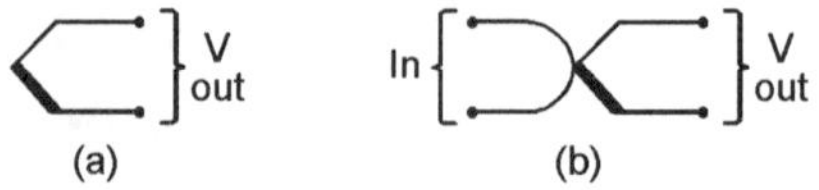

Figure 1.17 *Symbols of (a) a conventional and (b) an electrically-heated thermocouple device*

Thermocouples

When a junction is formed between two dissimilar metals a thermo-electric (temperature-dependent) voltage is generated across the junction. Thermocouples are devices in which the two types of metal are specially chosen to exploit this effect for temperature measurement purposes. A thermocouple using a copper and copper-nickel junction, for example, has a useful 'measurement' range from –100°C to +250°C, and has a typical sensitivity of 42μV per °C over the positive part of that range. Some thermocouples using other types of metal have useful measurement ranges that extend well above +1100°C. *Figure 1.17(a)* shows the symbol used to denote a normal thermocouple. In some special types of thermo-couple device the junction can be heated via a d.c. or r.f. current passed through a pair of input terminals; the thermocouple output can then be used to indicate the magnitude of the input current or power. Devices of this type use the symbol shown in *Figure 1.17(b).*

Light-dependent resistors (LDRs)

An LDR (also known as a cadmium sulphide (CdS) photocell) is a passive device with a resistance that varies with visible-light intensity. *Figure 1.18* shows the device's circuit symbol and basic construction, which consists of a pair of metal film contacts separated by a snake-like track of light-sensitive cadmium sulphide film; the structure is housed in a clear plastic or resin case. LDRs have many practical applications in security and auto-control systems.

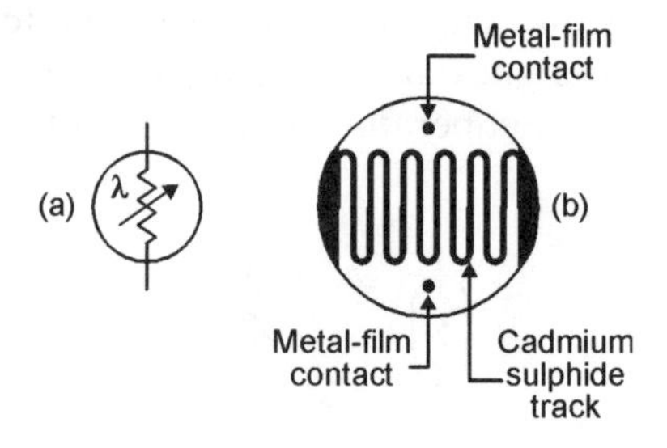

Figure 1.18 *LDR symbol (a) and basic structure (b)*

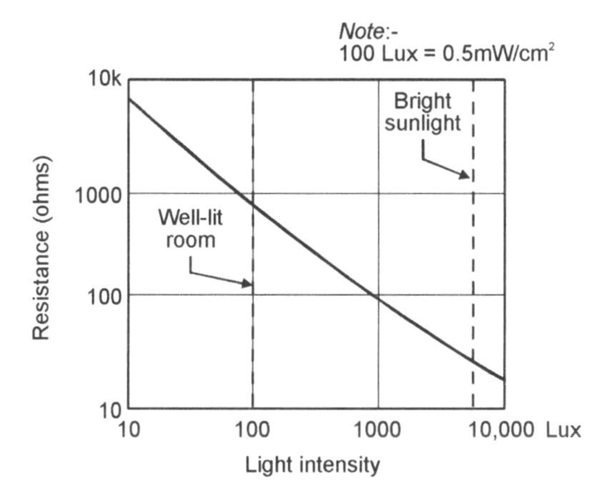

Figure 1.19 *Typical characteristic curve of an LDR with a 10mm face diameter*

Figure 1.19 shows the typical photoresistive graph that applies to an LDR with a face diameter of about 10mm; the resistance may be several megohms under dark conditions, falling to about 900R at a light intensity of 100 Lux (typical of a well lit room) or about 30R at 8000 Lux (typical of bright sunlight).

Microphones

Microphones are acoustic-to-electrical transducers and have a number of uses in eavesdropping and other security applications. The three best known types of electrical microphones are the moving-coil ('dynamic'), ribbon, and piezoelectric ('crystal') types. In most security electronics application, microphones are required to be small but sensitive types that generate medium-fidelity outputs; electronic 'electret' microphones are widely used in such applications.

Electronic sensor devices

An 'electronic' sensor may take the form of a single semiconductor component such as a photodiode or phototransistor, or may be a combination of electrical and/or electronic components that together perform a particular sensing function; examples of the latter type are electronic key-pad locks and light-beam alarms. The most important of such devices are described in this section.

Photodiodes

When p–n silicon junctions are reverse biased their leakage currents and impedances are inherently photo-sensitive; they act as very high impedances

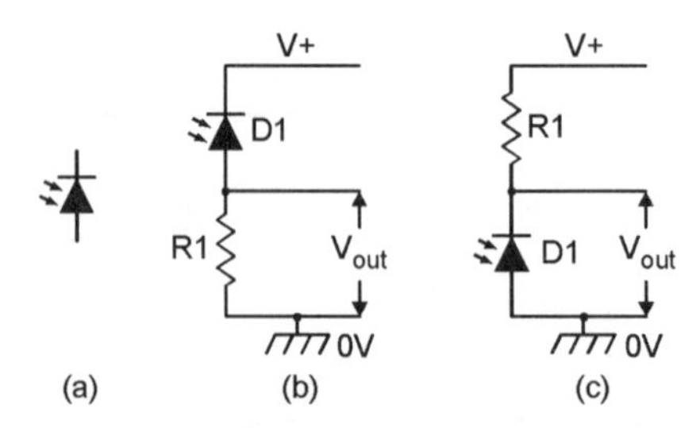

Figure 1.20 *Photodiode symbol (a) and alternative ways ((b) and (c)) of using a photodiode*

under dark conditions and as low impedances under bright ones. Normal diodes have their junctions shrouded in opaque material to inhibit this effect, but photodiodes are made to exploit it and use a translucent casing material; some photodiodes are made to respond to visible light, and some to infra-red (IR) light. *Figure 1.20(a)* shows the standard symbol of a photodiode. In use, the photodiode is simply reverse biased and the output voltage is taken from across a series resistor, which may be connected between the diode and ground as shown in *Figure 1.20(b)*, or between the diode and the positive supply line, as in *Figure 1.20(c)*.

Phototransistors

Ordinary silicon transistors are made from an npn or pnp sandwich, and thus inherently contain a pair of photo-sensitive junctions. Some types are available in phototransistor form, and use the standard symbol shown in *Figure 1.21(a)*. *Figures 1.21(b)* to *1.21(d)* show three basic ways of using a phototransistor; in each case the base–collector junction is effectively reverse biased and thus acts as a photodiode. In *(b)* the base is grounded, and the transistor acts as a simple photodiode. In *(c)* and *(d)* the base terminal is

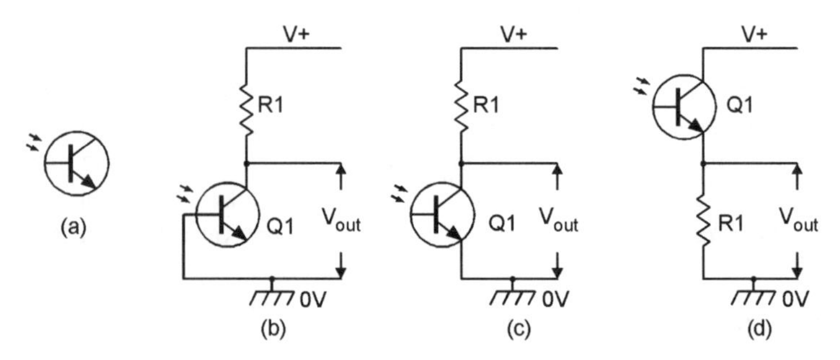

Figure 1.21 *Phototransistor symbol (a) and alternative ways ((b) to (d)) of using a phototransistor*

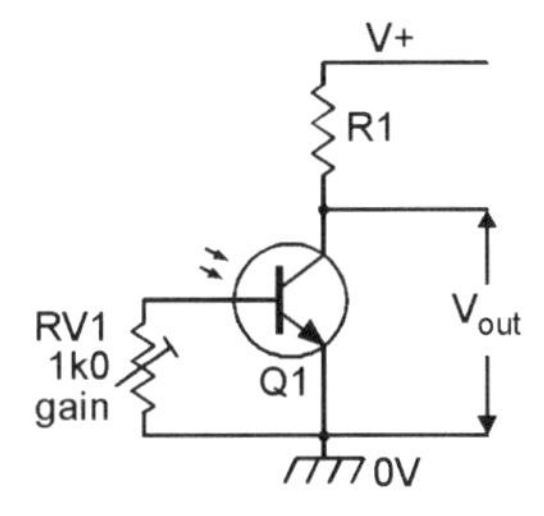

Figure 1.22 *Variable-sensitivity phototransistor circuit*

open circuit and the photo-generated currents effectively feed directly into the base and, by normal transistor action, generate a greatly amplified collector-to-emitter current that produces an output voltage across series resistor R1.

The sensitivity of a phototransistor is typically one hundred times greater than that of a photodiode, but its useful maximum operating frequency (a few hundred kHz) is proportionally lower than that of a photodiode (tens of MHz). Some phototransistors are made in very-high-gain Darlington form. A phototransistor's sensitivity (and operating speed) can be made variable by wiring a variable resistor between its base and emitter, as shown in *Figure 1.22*; with RV1 open circuit, phototransistor operation is obtained; with RV1 short circuit, photodiode operation occurs.

Optocouplers

An optocoupler is a device housing a LED (usually an IR type) and a matching phototransistor; the two devices are optocoupled but are electrically isolated from each other and – in a normal type of optocoupler – are mounted in a light-excluding housing. *Figure 1.23* shows a basic optocoupler 'usage' circuit. The LED is used as the input side of the circuit, and the phototransistor as the output. Normally, SW1 is open and the LED and Q1

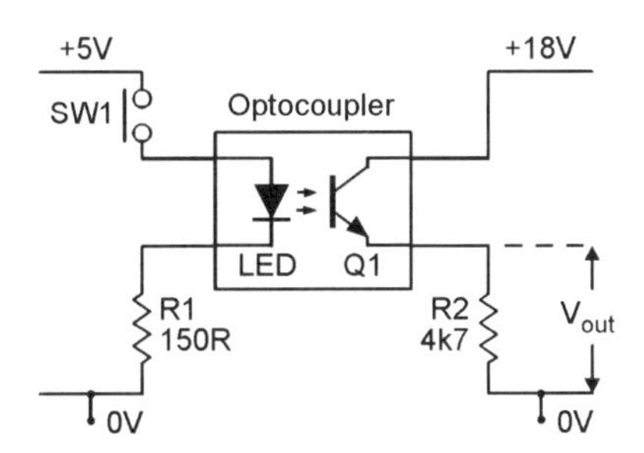

Figure 1.23 *Basic optocoupler circuit*

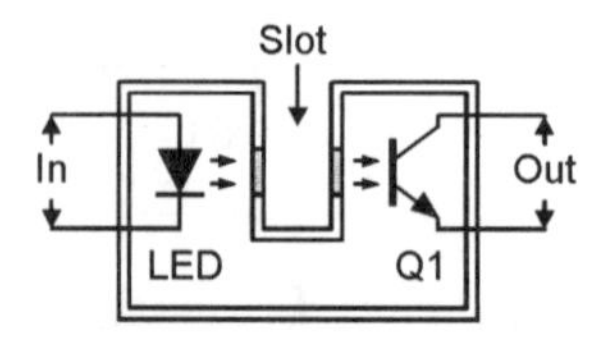

Figure 1.24 *Slotted optocoupler device*

are thus off. When S1 is closed a current flows through the LED via R1, and Q1 is turned on optically and generates an output voltage across R2. The output circuit is thus controlled by the input one, but the two circuits are fully isolated electrically ('isolation' is the major feature of this type of optocoupler, which can be used to couple either digital or analogue signals).

The *Figure 1.23* device is a standard type of optocoupler. There are, however, two special types of optocoupler that are of particular value in security electronics applications, and these are shown in *Figures 1.24* and *1.25*. The *Figure 1.24* 'slotted' device has a slot moulded into the package between the LED light source and the Q1 light sensor. Light can normally pass from the LED to Q1 via a pair of windows in the slot walls, but can be blocked by placing an opaque object in the slot. The slotted optocoupler can thus be used in a variety of 'presence detecting' applications, such as limit switching and dark-liquid level detection.

The *Figure 1.25* 'reflective' optocoupler has the LED and the Q1–Q2 Darlington light sensor optically screened from each other within the package but arranged so that they both point outwards – via windows – towards an external point. The construction is such that an optocoupled link can be set up by a reflective object (such as metallic paint or tape) placed a short distance outside the package, in line with the LED and Q1. The reflective optocoupler can thus be used in applications such as tape-position detection, engine- or motor-shaft RPM measurement, or marked-object theft (illegal movement) detection, etc.

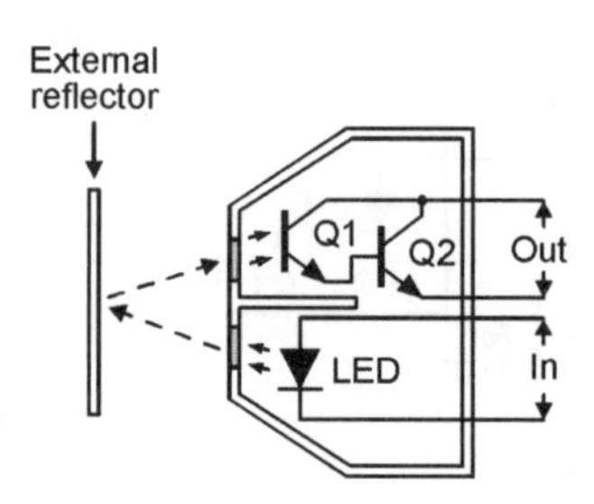

Figure 1.25 *Reflective optocoupler*

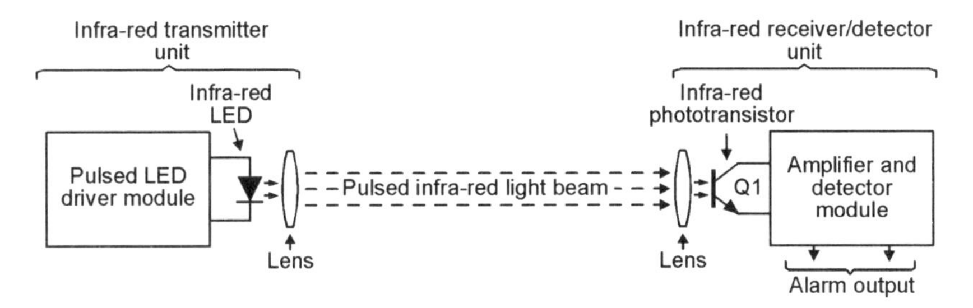

Figure 1.26 *Basic infra-red light-beam alarm system*

Light-beam units

Most modern 'light-beam' units work on the basic principle illustrated in *Figure 1.26*, in which a focused invisible beam of pulsed infra-red light is generated by a transmitter unit, and is detected at a remote point by a matching lens and receiver/detector unit. Normally, the unit is configured so that the receiver generates an alarm output if the IR beam is interrupted. Such units have useful operating ranges of up to thirty metres and are often used in industry in automatic batch counting and safety-switch operating applications, and in commercial and domestic applications as intruder-detecting security alarms. Simple single-beam alarms of the basic *Figure 1.26* type have fairly low values of reliability, since they can easily be triggered by insects settling on one or other of the unit's lenses, but dual-beam types of alarm – in which both the transmitter and the receiver use two lenses placed a few inches apart – have high values of reliability.

Pyroelectric IR detectors

Some special crystals and ceramics generate electric charges when subjected to thermal variations or uneven heating; this is known as a pyroelectric effect. Pyroelectric infra-red detectors incorporate one or two elements of this type, plus a simple filtering lens and a field-effect transistor (FET), configured in the basic way shown in *Figure 1.27(a)*. The basic action of the device is such that – if a human body moves within the visual field of its pyroelectric elements – part of the radiated infra-red energy of that body falls on the surface of the elements and is converted into a minute variation in surface temperature and a corresponding variation in the element's output voltage. When the unit is wired as shown in the *Figure 1.27(b)* basic usage circuit, this movement-inspired voltage variation is made externally available via the buffering JFET and capacitor C1 and can, when suitably amplified and filtered, be used to activate an alarm when a human body movement is detected.

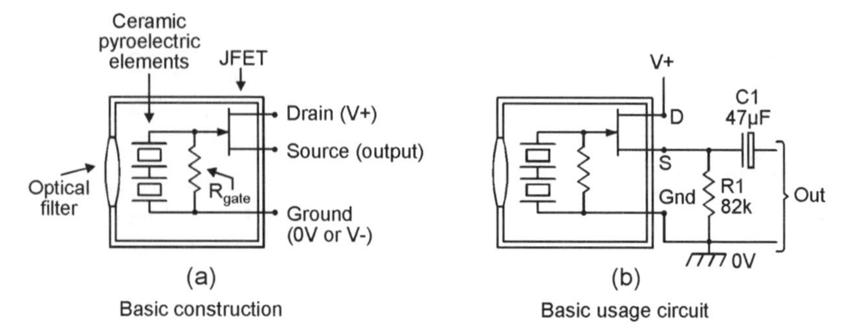

Figure 1.27 *Basic construction (a) and usage circuit (b) of a pyroelectric infra-red detector*

Note that pyroelectric IR detector circuits of the basic type described above have, because of the small size of the detector's light-gathering lens, maximum useful detection ranges of just over one metre, but that this range can be extended to more than ten metres with the aid of a relatively large external light-gathering/focusing lens of the type used in modern passive infra-red (PIR) movement detector systems (see *Figure 1.4* and its associated text).

Piezoelectric transducers

A piezoelectric transducer is an electro-constrictive device that converts a varying electrical signal into a sympathetic set of fine mechanical variations, or vice versa. Devices of this type include piezo sounders, 'crystal' earphones and microphones, ordinary quartz crystals, and ultrasonic transducers. Most devices of the later type are sharply tuned low-power units designed to peak at about 40kHz, and are supplied in matching pairs, with one optimized for use as a signal transmitter and the other as a signal receiver. They are useful in many remote control and distance-measurement applications, and in 'doppler effect' intruder alarm systems of the basic type shown in *Figure 1.28*.

The *Figure 1.28* intruder alarm system consists of three main elements. The first is a transmitter (Tx) that floods the room with 40kHz ultrasonic signals, which bounce back and forth around the room. The second is a receiver (Rx) that picks up and amplifies the reflected signals and passes them to a phase comparator, where they are compared with the original 40kHz signal. If nothing is moving in the room the Tx and Rx signal frequencies will be the same, but if an object (an intruder) is moving in the room the Rx signal is doppler-shifted by an amount proportional to the rate of object movement (by about 66Hz at 10 inches/sec). The l.f. output of the comparator is passed on to the third system element, the alarm activator, which is a signal conditioner that rejects spurious and out-of-limits signals, etc., and activates the

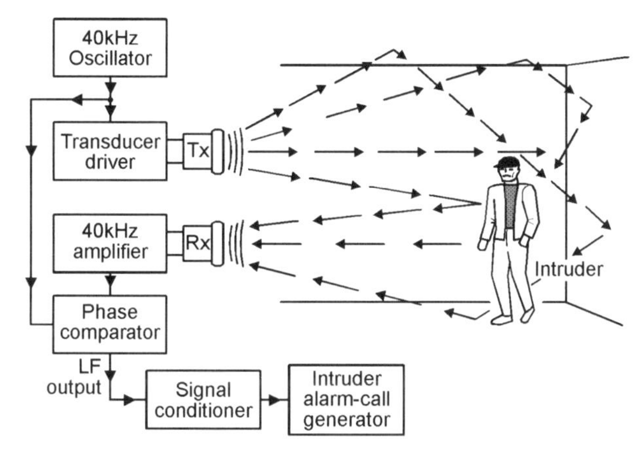

Figure 1.28 *Block diagram of an ultrasonic (40kHz) doppler-effect intruder alarm system*

alarm-call generator only if an intruder is reckoned to be genuinely present. In practice, many systems of this basic type have poor reliability when set to high-sensitivity levels, since they can easily be false-triggered by draughts, central-heating air currents, and curtain movements, etc. Low-sensitivity versions of the system are often used to protect small areas, such as the interiors of automobiles, however, and usually have high values of reliability.

Electret microphones

Electret microphones are modern highly efficient 'capacitor' microphones, and use the basic form of construction shown in *Figure 1.29*. Here, a lightweight metallized diaphragm forms one plate of a capacitor, and the other plate is fixed and is metallized on to the back of a slab of insulating material known as electret; the capacitance value thus varies in sympathy with the

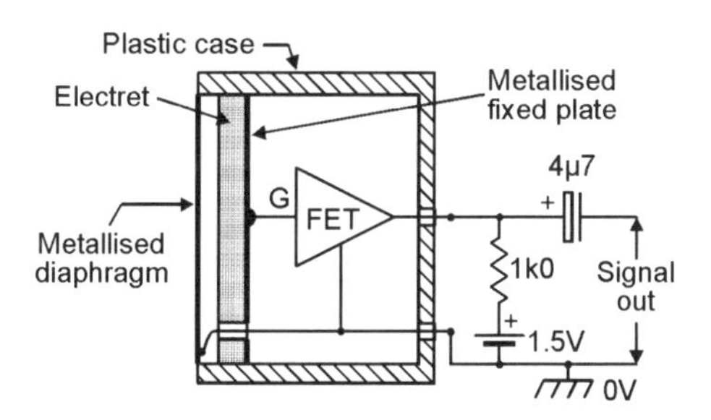

Figure 1.29 *Basic elements of a modern electret microphone*

applied acoustic (sound) signal. The electret material holds a fixed electrostatic charge that is built in during manufacture and can be held for an estimated 100-plus years; this charge is applied between the two plates. The voltage across the capacitor equals this charge divided by the capacitance value and – since this varies in sympathy with the applied acoustic signal – varies in sympathy with the acoustic signal. This signal is fed to the outside world via a built-in IGFET transistor, which needs to be powered externally from a battery (1.5V to about 9V) via a 1k0 resistor, as shown. Electret microphones are robust and inexpensive and give a good performance up to about 10kHz; they are useful in many audio sound pick-up applications, particularly in sound-activated alarms and eavesdropping units.

Key-pad switches

These are modern and greatly superior replacements for conventional electromechanical key-switches, and are opened by typing a secret multi-digit code number into a simple key-pad, rather than by the use of an easily lost or stolen mechanical key. Typically, units of this type take the basic form shown in *Figure 1.30*, in which the key-pad houses twelve push-button switches, notated with the numerals 0 to 9 and the letters C (change code) and D (disable/enable), plus two state-indicating LEDs. The switches are (in this example) arranged in four vertical and three horizontal columns, which are wired to a clocked decoder and control logic network that converts each digit keystroke into a 4-bit binary code and compares it with the 4-bit code that is stored in the matching line of the system's RAM; if the entire code number (which is usually 4 to 8 digits long) is typed in without error, the switch opens and performs a useful function (opens a door or gives access to an engine's start-up system, etc.), but if the correct code is not entered within three attempts the lock automatically goes into a time-controlled shutdown or alarm mode.

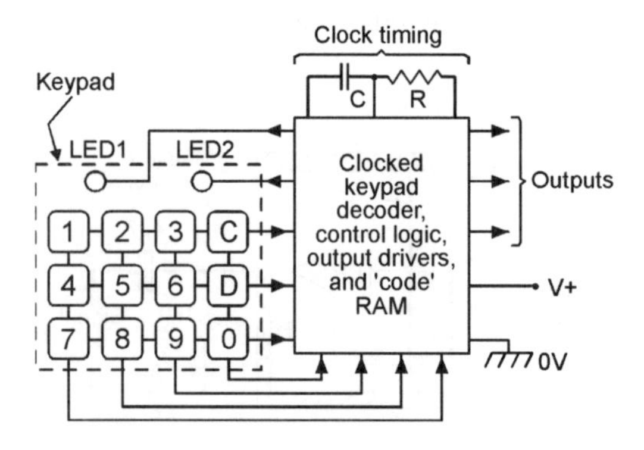

Figure 1.30 *Basic elements of a key-pad security switch*

In the above system, the secret code number can be changed at any time by simply typing in the existing code, pressing the 'C' switch once (to gain direct access to the RAM), typing in the new code number, and then pressing the 'C' switch again (to return to normal operation). The entire keyswitch can be disabled (for a time-controlled period) or re-enabled at any time by operating the 'D' switch, which gives a toggling disable/enable type of action; the key-pad switch's operating mode is displayed at all times via the two state-indicating LEDs.

Digital time switches

Figure 1.31 shows a symbolic representation of a digital time-operated SPST electric switch, in which the switch arm is controlled via accurately timed digital circuitry and can be programmed to turn on and off at any desired times of the day or week. Digital time switches offer far greater precision than normal analogue types, and are used in many light-switching and solenoid-operating security applications.

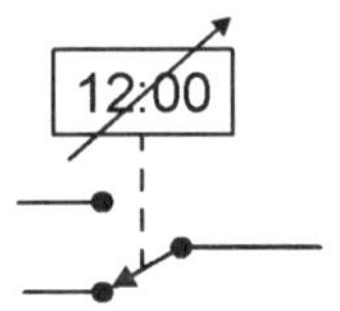

Figure 1.31 *Symbolic representation of a digital time-operated SPST electric switch*

Miscellaneous electronic sensors

A variety of special-purpose electronic sensors of value in security applications but not so-far mentioned in this section are also available from some specialist dealers. Among the most useful of these are radioactive 'smoke detector' elements that respond to various ionized particles, humidity sensors, strain gauges, hall-effect devices that respond to magnetic field strength (flux density), and 'gas' sensors that react to gases such as propane, butane, methane, isobutane, petroleum gas, natural gas, and 'town' gas. A few of these devices are described in some detail in later chapters of this volume.

Data links

Data links are (apart from the actual signal processing unit) one of the three major elements of any electronic security system, the other two elements being the sensing unit(s) and the response unit. All practical security systems use at least two data links (see *Figure 1.1*), which may have individual lengths ranging from less than one millimetre to many thousands of kilometres, depending on the specific application. Most data links fit into one or other of three basic types, being either hard-wired types, optocoupled types, or wireless types, as described in the rest of this *Data links* section.

Hard-wired data links

Most hard-wired data links take the form of a length of multi-cored cable, used to link a sensor or response unit to the alarm system's main control unit. *Figures 1.32* to *1.34* show examples of such cables used to link a sensor switch to the input of a simple burglar alarm unit. In *Figure 1.32* the sensor switch is a normally-closed one of the type used to protect doors or windows and is connected to the unit via a 2-wire (or 2-core) data link; this circuit's basic action is such that – when key-switch S2 is closed – Q1 and the alarm both turn on if sensor switch S1 is opened or the data link is accidentally or deliberately cut; this circuit thus has an inherently good anti-tamper performance.

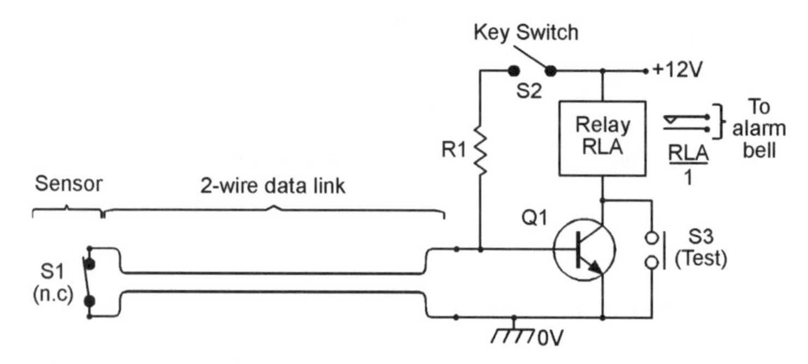

Figure 1.32 *2-wire data link used with a normally-closed sensor switch offers good security*

In the *Figure 1.33* circuit the sensor switch is a normally-open type such as a pressure mat switch, and is connected to the unit via a 2-wire data link; this circuits basic action is such that – when key-switch S2 is closed – Q1 and the alarm normally both turn on if sensor switch S1 is closed, but will fail to operate if the data link is accidentally or deliberately cut; this circuit thus has a poor anti-tamper performance.

Finally, *Figure 1.34* shows a high-security version of the above circuit. In this case the sensor switch is again a normally-open type such as a pressure mat switch, but is connected to the unit via a data link that uses four wires,

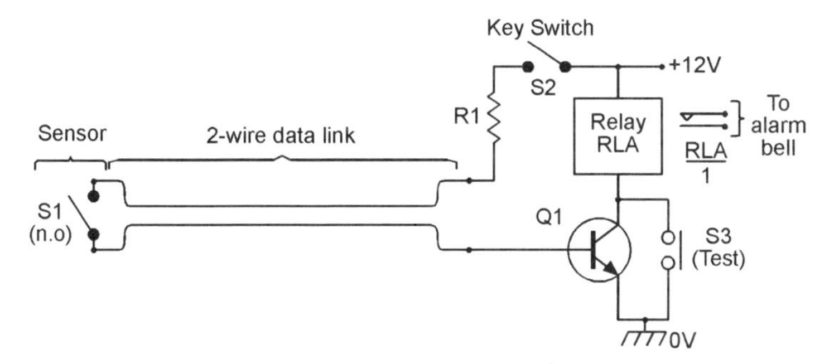

Figure 1.33 *2-wire data link used with a normally-open sensor switch offers very poor security*

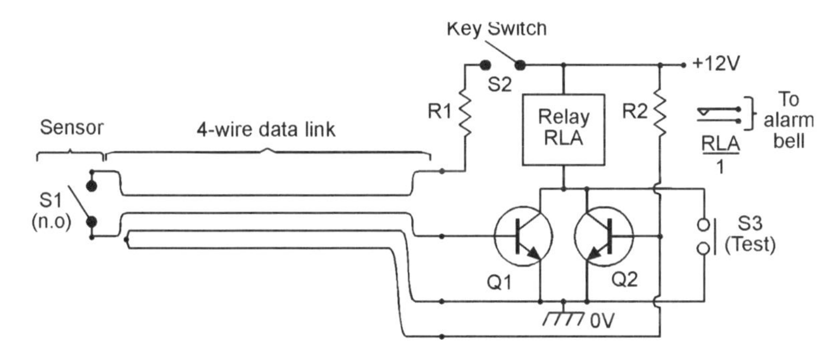

Figure 1.34 *4-wire data link used with a normally-open sensor switch offers excellent security*

two of which serve an anti-tamper function; this circuit's basic action is such that the alarm normally turns on via R1–Q1 if sensor switch S1 and key-switch S2 are both closed, but operates instantly (even if key-switch S2 is open) via Q2–R2 if the 4-wire data link is accidentally or deliberately severed; this circuit thus has an excellent anti-tamper performance and is often used in department stores and other places in which the public have easy access to parts of the alarm system.

Optocoupled data links

Optocoupled data links are often used in applications where it is not possible or convenient to use a hard-wired data link, and come in three basic types, being either infra-red 'light-beam' types, fibre optic 'light guide' types, or laser beam types. Light-beam types are used mainly in short-range (less than 6 metres) remote control applications, but can – if used with a good lens system – be effective at ranges up to about 20 metres; units of the latter type are sometimes used (in domestic applications) as a data link between a shed or other remote building's intrusion sensor and a main alarm unit. Fibre optic light guide data links are used mainly in applications where the link is fairly long (greater than ten metres) and needs a wide signal bandwidth. Laser beam data links are used mainly in medium-range applications in which a hard-wired link can not be used; they are sometimes used (illegally) in remote eavesdropping applications, in which the beam is bounced off the window of a room in which a secret conversation is taking place, the return beam being modulated by the window's acoustic pick-up signals.

Wireless data links

Wireless data links – usually operating at 418MHz or 458MHz – are widely used in modern domestic burglar alarm systems (see *Figure 1.5*) to link the system's various sensors to the main control unit, thus greatly easing installation problems and enabling the system to be remote-controlled via a small key-fob signal transmitter. Most systems of this type have typical control ranges of up to 30 metres, but some sophisticated systems can be interfaced with both the domestic heating control unit and with the normal telephone system, enabling alarm and heating systems to be remotely monitored or controlled over a range of thousands of miles. The owner of such a system can, while holidaying or working abroad, use a fixed or mobile phone to check the home's security at any time, or can use it to remotely turn on the building's central heating system prior to eventually returning home.

Alarm response units

Alarm response units are the final major elements in any electronic security system, and usually take the form of a simple relay, some type of electromagnet, a solenoid- or motor-operated mechanism, or (in burglar alarm and other high-level security systems) a sound generator and/or a light strobe unit; brief details of units of these types are given in the rest of this *Alarm response units* section.

Relay response units

Relays are electrically operated switches that can be used to activate virtually any external electrical devices (such as lamps, sirens, motors, etc.). Relays come in two basic types, one being the 'reed' type that has already been shown in *Figure 1.12(a)* and the other being the conventional electromagnetic type that takes the basic form shown in *Figure 1.35*. Here, a multiturn coil is wound on an iron core to form an electromagnet that can move an iron lever or armature which in turn can close or open one or more sets

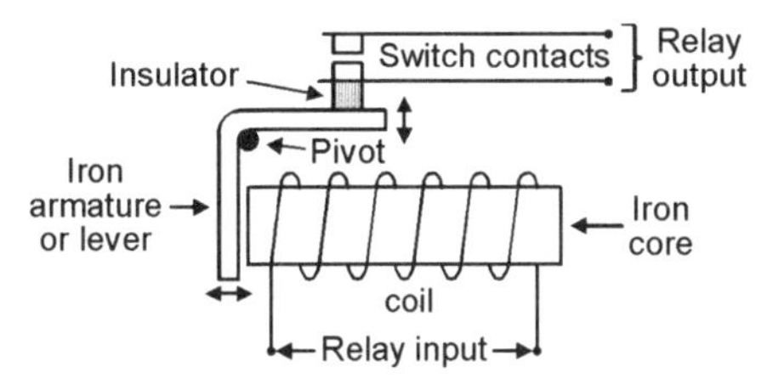

Figure 1.35 *Basic design of a standard electromagnetic relay*

of switch contacts. The operating coil (which requires only a modest operating current) is electrically fully isolated from the switch contacts (which can control fairly high currents), and can be shown as separate elements in circuit diagrams, as shown in *Figure 1.36*, which represents a relay with a 12V, 120R coil and a single set of normally-open (n.o.) switch contacts.

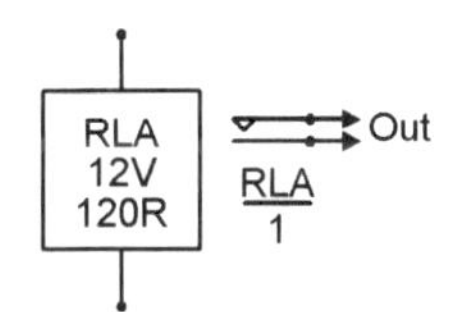

Figure 1.36 *Representation of a 12V, 120R relay with one set of n.o contacts*

Relays with a single set of n.o contacts are usually used in the basic non-latching mode shown in *Figure 1.37(a)*, in which the relay closes when S1 is closed and opens when S1 is opened. Relays with two (or more) sets of n.o. contacts can also be used in the self-latching mode shown in *Figure 1.37(b)*, in which n.o. contacts RLA/2 are wired in parallel with S1 so that they close and lock (latch) the relay on as soon as S1 is closed; once the relay has locked on it can be turned off again by briefly breaking the supply connections to the relay coil.

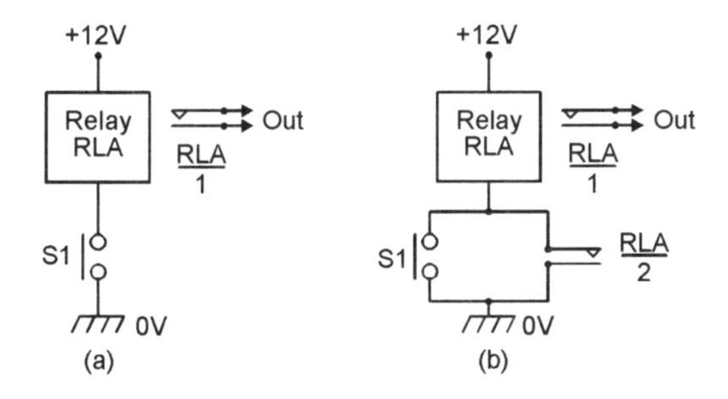

Figure 1.37 *Relay switch used in (a) non-latching or (b) self-latching modes*

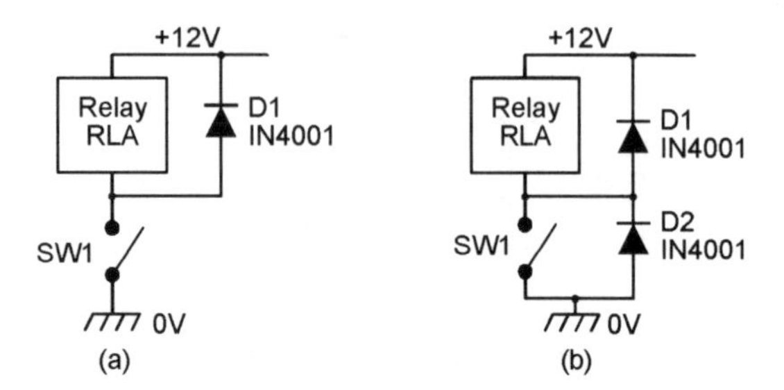

Figure 1.38 *Relay coil using (a) single-diode or (b) two-diode coil damper*

Relay coils are highly inductive and may generate back-emfs of hundreds of volts if their coil currents are suddenly interrupted. These back-emfs can easily damage switch contact or solid-state devices connected to the coil, and it is thus often necessary to 'damp' them via protective diodes, as shown in *Figures 1.38*. In *Figure 1.38(a)* the coil damping is provided via D1, which prevents switch-off back-emfs from driving the RLA-SW1 junction more than 600mV above the positive supply line. This form of protection is adequate for normal switching applications. In *Figure 1.38(b)* the damping is provided via two diodes that stop the RLA–SW1 junction swinging more than 600mV above the positive supply rail or below the zero-volts rail. This form of protection is recommended for all applications in which SW1 is replaced by a transistor or other solid-state switching device.

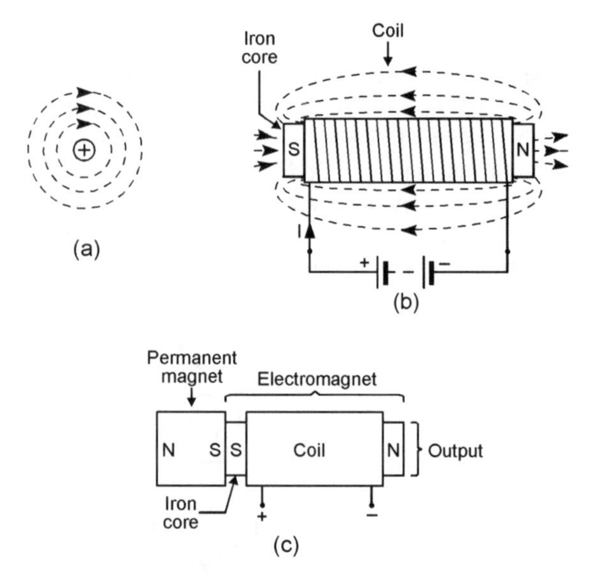

Figure 1.39 *Diagrams showing the magnetic fields generated by (a) a current-carrying wire and (b) an electromagnet, and the basic construction of (c) an energize-to-release type of holding magnet*

Electromagnet units

Electromagnet units are widely used in industrial and commercial applications to control the hold or release actions of security doors and safety guards and gates, etc. *Figure 1.39* illustrates basic electromagnet operating principles. When a current is passed through a wire, a magnetic field is generated about the axis of the wire, as shown in the cross-sectional view in *Figure 1.39(a)*. When such a wire is wound as a coil on an iron-cored former, the fields of the individual turns interact in the way shown in *Figure 1.39(b)*, causing the core to act like a normal bar magnet (with north and south poles) when the coil is energized, but to act like a piece of non-magnetic iron when the coil is not energized. This basic type of electromagnet is thus used in energize-to-hold applications.

Figure 1.39(c) shows a useful variant of the normal electromagnet. Here, a permanent magnet is fixed to one end of the electromagnet's iron core, in the polarity shown in the diagram, and the other end of the core forms the output of the unit. When the electromagnet is not energized, its iron core acts as a simple extension of the permanent magnet, with its output acting as the southern pole of the magnet, but when the electromagnet *is* energized its magnetic field opposes that of the permanent magnet, and (if the two opposing fields are of equal strength) the unit's output is thus demagnetized. This type of unit thus acts as an energize-to-release type of holding magnet.

Solenoid-operated units

Solenoids are electromagnetic devices that are designed to move an iron ram or an armature and thereby activate a device such as a power switch, a safety latch, or a control valve or tap, etc. They consist of a multi-turn coil that is wound about the axis of a fixed or moving iron core. Fixed-core types act as simple electromagnets that move an external iron armature when the coil is energized; the best known example of this type of unit is the standard electromagnetic relay shown in *Figure 1.35*. In moving-core types of solenoid the coil is wound on a plastic or waxed-paper tube in which the iron core (which usually takes the form of a ram) is free to move; the basic action of this type of unit is such that the centre-of-mass of the iron core (ram) is forced into a central position within the coil when the coil is energized, but may be forced into a different position (via a spring, etc.) when the coil is not energized.

Moving-core solenoids come in several basic variants, and the three most widely used of these are shown in *Figure 1.40*. The most widely used type gives a simple linear movement of the ram, as shown in *Figure 1.40(a)*. Here, the ram is normally biased to the left-of-centre of the coil by two collars and a spring, but is forced to the right (to the coil's central position) when the coil is energized, thus giving a thrust action at the right-hand end of the ram

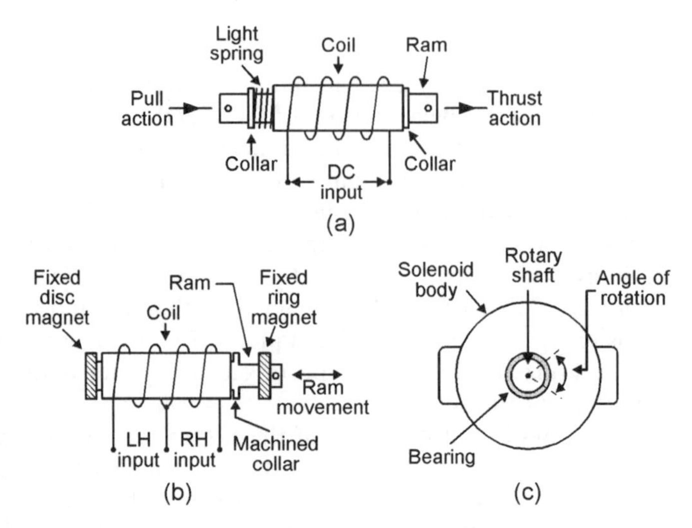

Figure 1.40 *Simplified diagrams of (a) a conventional moving-core linear solenoid, (b) a magnetically latching split-coil moving-core linear solenoid, and (c) a rotary solenoid*

and a pull action at the left-hand end; many practical solenoids of this basic type are designed to give only a thrust action or only a pull action.

A useful variant of the moving-core linear solenoid is the magnetically latching split-coil type shown in *Figure 1.40(b)*. Here, when a pulse of energizing current is fed to the right-hand (RH) side of the split coil, the ram is forced to the right until a machined collar makes contact with a fixed ring magnet, which latches the ram in that position when the coil is de-energized. Once the ram has latched into this position it can only be unlatched by feeding a pulse of energizing current to the left-hand (LH) side of the coil, thus forcing the ram to the left until its left face makes contact with a fixed disc magnet, which latches the ram into this alternative position, and so on. Magnetically latching solenoids are useful where low mean power consumption is required. Note that simple split-coil solenoids are widely used as points-controllers in model railway systems, but do not incorporate magnetic latching.

Finally, the third type of moving-core solenoid is the rotary movement type shown in *Figure 1.40(c)*, in which the solenoid's linear action is converted into rotary form via a simple crank or link mechanism. These units typically give maximum shaft rotation angles in the range 45° to 95°.

Bells and buzzers

Electric bells and buzzers are widely used sound-generating alarm response units; *Figure 1.41* shows their typical basic construction and electrical equiv-

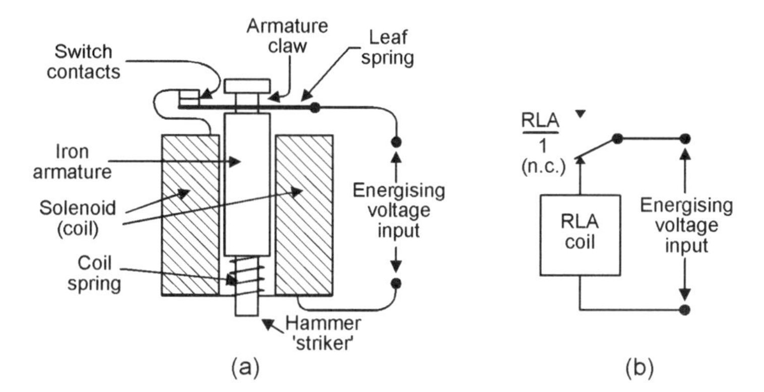

Figure 1.41 *Typical basic construction (a) and equivalent circuit (b) of an electric bell or buzzer*

alent circuit. They consist of an iron armature that can move freely within a solenoid that can be energized via a pair of normally-closed switch contacts and a leaf spring. Normally, the armature is forced out of the solenoid by a light coil spring. When a suitable energizing voltage is connected to the circuit the solenoid pulls the armature downwards until its hammer striker hits a sounding board (in a buzzer) or metal dome (in a bell); at this point a claw at the other end of the armature pulls open the switch contact via the leaf spring, and the armature shoots outwards again under the pressure of the coil spring until the switch contacts close again and the process then repeats *ad infinitum*. Electric bells and buzzers are thus self-interrupting inductive devices; electric bells have their acoustic output energy concentrated into a narrow 'tone' band and are thus reasonably efficient, but electric buzzers generate a broad 'splurge' of sound and are very inefficient.

Motor-operated units

Electric motors are widely used in industry and commerce to give automatic operation of safety and security doors, and to automatically operate customer-access doors and gates under approved safety/security conditions.

Sound-generator/light-strobe units

All emergency-warning security and safety systems should (ideally) be fitted with an efficient attention-grabbing sound-generator system, to warn all and sundry of the existence of the emergency state, and with some form of light-strobe unit, to visually indicate the precise source of the emergency signals. In buildings, the sound generator may take the form of an electromechanical alarm bell or a piezoelectric or horn-speaker-based electronic siren, and the visual warning may come from a special light-strobe; in automobiles, the sound generator may take the form of a siren or a unit that pulses the vehicles horn, and the visual warning should be obtained by flashing the vehicles lights. In all cases, the alarm-condition indicator unit must be fitted with an automatic timing mechanism that shuts it down after a pre-set period (typically less than 15 minutes) of operation.

In burglar alarm systems, the sound-generator and light-strobe units should be fitted together in a special alarm box and mounted high up on an external wall that (ideally) faces onto a well-used street or passageway. The box should have a built-in back-up battery that is charged via the system's control panel cables, and the unit should automatically activate the alarm if this cable is cut; the alarm box should be fitted with some form of microswitch that automatically activates the alarm if any attempt is made to open its front cover or pull it from the wall. Units of this type are readily available from electronic alarm system suppliers.

2

Contact-operated security circuits

Contact-operated security circuits are units that are activated by the opening or closing of a set of electrical contacts. These contacts may take the form of a simple push-button switch, a pressure pad switch, or a magnetically activated reed switch, etc. The security circuit's output may take the form of some type of alarm-sound generator, or may take the form of a relay that can activate any external electrical device, and may be designed to give non-latching, self-latching, or one-shot output operation.

Contact-operated security systems have many practical applications in the home, in commercial buildings, and in industry. They can be used to attract attention when someone operates a push switch, or to give a warning when someone opens a door or treads on a pressure pad or tries to steal an item that is wired into a security loop, or to give some type of alarm or safety action when a piece of machinery moves beyond a preset limit and activates a microswitch, etc. A wide range of practical contact-operated security circuits are described in this chapter.

Bell and relay-output circuits

Close-to-operate circuits

The simplest type of contact-operated security circuit consists of an alarm bell (or a buzzer or electronic 'siren-sound' generator, etc.) wired in series with a normally-open (n.o.) close-to-operate switch, the combination being wired across a suitable battery supply, as shown in the basic 'door-bell' alarm circuit of *Figure 2.1*. Note that any desired number of n.o. switches can be wired in parallel, so that the alarm operates when any of these switches are closed. This type of circuit gives an inherently non-latching type of operation, and has the great advantage of drawing zero standby current from its supply battery.

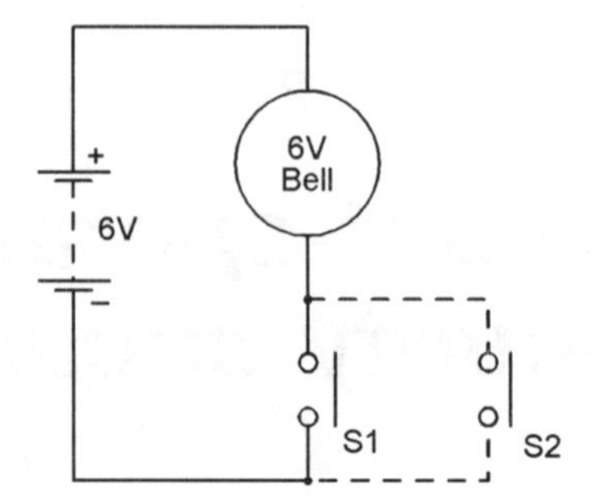

Figure 2.1 *Simple door-bell type close-to-operate alarm circuit*

A disadvantage of the basic *Figure 2.1* circuit is that it passes the full 'alarm' current through the n.o. operating switches and their wiring, so the switches must be fairly robust types, and the wiring must be kept fairly short if excessive wiring volt-drops are to be avoided. This latter point is of particular importance in security applications in which the circuit is used with several widely separated n.o. switches. The solution to this problem is to activate the bell via a 'slave' device (which is fitted close to the bell but requires a fairly low input current), and to activate this slave device (and thus the bell) via the security switches. *Figures 2.2* to *2.6* show a variety of such circuits, in which the slave device takes the form of a relay, a power transistor, or an SCR.

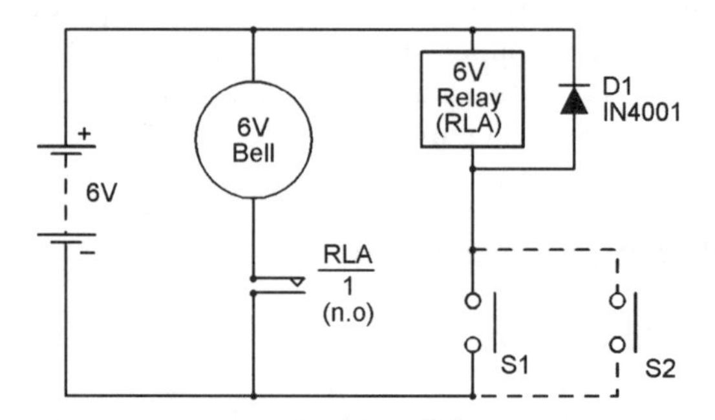

Figure 2.2 *Relay-aided non-latching close-to-operate alarm*

Figure 2.2 shows a relay-aided version of the close-to-operate alarm circuit. Here, the parallel-connected n.o. switches are wired in series with the coil of a 6V relay (which typically draws an operating current of less than 100mA), and the relay contacts (which can typically switch currents of several amps) are wired in series with the alarm bell, and both combinations are wired across the same 6V supply. Thus, when the switches are open the relay is off and its contacts are open, so the bell is off, but when any one or more of the switches is closed the relay is driven on and its contacts close and activate the alarm bell. Note in the latter case that the switches and their wiring pass a current equal to that of the relay coil; the switches can thus be fairly delicate ones, such as sensitive reed types, and the wiring can be reasonably long. Silicon diode D1 is wired across the relay's coil to protect the switches against damage from the coil's switch-off back-emf.

The *Figure 2.2* circuit gives a non-latching form of operation, in which the alarm operates only while one or more of the operating switches is closed. In most high-security applications, the circuit should be a self-latching type in which the relay and alarm automatically lock on as soon as any one of the n.o. switches is closed, and can only be de-activated via a security key. *Figure 2.3* shows the above circuit modified to give this type of operation. Here, the relay has two sets of n.o. contacts, and one of these is wired in parallel with the n.o. switches so that the relay self-latches as soon as it is operated, and the entire circuit can be enabled or disabled/de-activated via key-switch S1, which is wired in series with the battery supply line. Circuits of this basic type are usually used in low-cost 'zone protection' applications, in which the 'zone' is a large room or shop floor, the S1 key-switch is located outside of the zone, and the n.o. trigger switches are hidden pressure mat switches or door- or window-operated microswitches fitted within the protected zone.

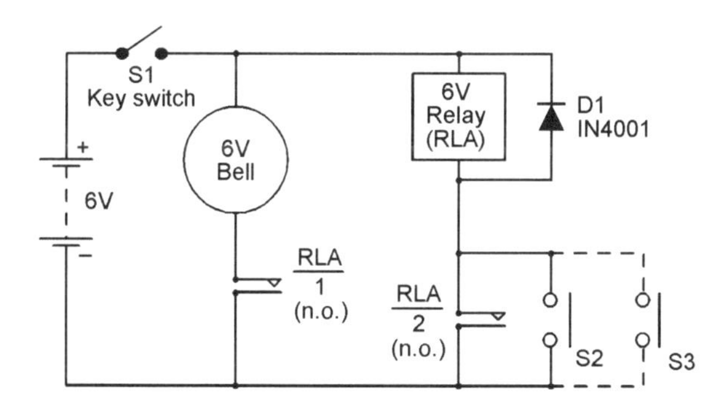

Figure 2.3 *Relay-aided self-latching close-to-operate security alarm*

An alternative solution to the *Figure 2.1* switch-and-wiring 'current' problem – but which can only be used in non-latching applications – is shown in *Figure 2.4*, in which npn power transistor Q1 is used as the slave device. Resistor R1 ensures that when any of the activating switches are closed, Q1's drive current is limited to less than 60mA, which (assuming that Q1 has a nominal current gain of at least ×25) enables the transistor to switch at least 1.5A through the alarm bell.

Another solution to the 'current' problem is to use an SCR (silicon controlled rectifier) as the slave device, as shown in *Figures 2.5* and *2.6*.

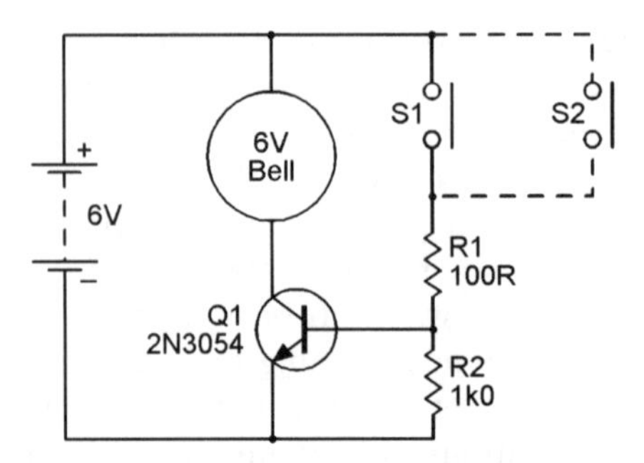

Figure 2.4 *Transistor-aided non-latching close-to-operate alarm*

These circuits rely on the fact that ordinary electromagnetic alarm bells are self-interrupting solenoid devices that incorporate a self-activating on/off switch in series with the solenoid's supply line. This switch is normally closed, allowing current to reach the solenoid and throw out a striker that hits the bell dome and simultaneously opens the switch, thus breaking the current feed and causing the striker to fall back again until the switch closes again, at which point the whole process starts to repeat, and so on; the bell's operating current is thus drawn in pulsed form.

In the *Figure 2.5* circuit the alarm bell is wired in series with an SCR that has its gate current derived from the positive supply line via current-limiting resistor R1 and via the parallel-connected n.o. security switches, which (when R1 has a value of 1k0) pass operating currents of only a few milliamps. When all the switches are open, the SCR and alarm bell are off, but when any one of the switches is closed it feeds gate current to the SCR via R1, so the SCR turns on and activates the bell. Note in this design that, since the bell is a self-interrupting device, the circuit effectively gives a non-latching type of operation in which the SCR and bell only operate while one or more of the switches are closed.

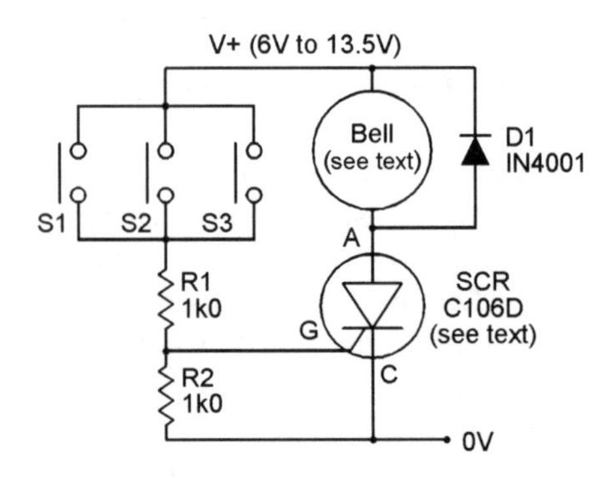

Figure 2.5 *SCR-aided non-latching close-to-operate alarm*

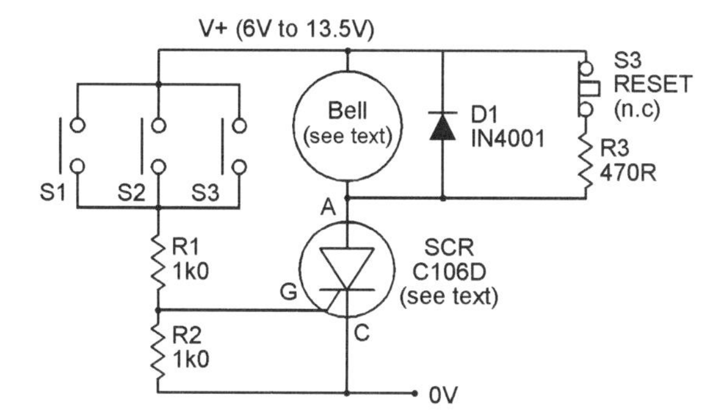

Figure 2.6 *SCR-aided self-latching close-to-operate alarm*

Figure 2.6 shows how the above circuit can be modified to give self-latching operation. SCRs are inherently self-latching devices that, once they have been initially turned on, remain on until their anode current falls below a 'minimum holding' value, at which point the SCR unlatches and turns off. In the *Figure 2.5* circuit the SCR thus automatically unlatches each time the alarm bell self-interrupts, but in the modified *Figure 2.6* design the bell is shunted via R3, which is wired in series with n.c. switch S4, which ensure that the SCR's anode current does not fall below the C106's minimum holding current value when the bell self-interrupts, thus providing the circuit with a self-latching action.

Note that the C106 SCR used in the *Figure 2.5* and *2.6* circuits has an anode current rating of only 2A, so the alarm bell must be selected with this point in mind. Alternatively, SCRs with higher current ratings can be used in place of the C106, but this modification will probably necessitate changes in the R1 and R3 values of the circuits. Also note in these SCR circuits that, to compensate for the SCR's typical 1V anode-to-cathode volt drop, the supply voltage must be at least 1V greater than the nominal operating voltage of the alarm bell.

Open-to-operate circuits

A major weakness of the *Figure 2.1* to *2.6* circuits is that they do not give a 'fail-safe' form of operation, and give no indication of a fault condition if a break occurs in the contact-switch wiring. This snag is overcome in circuits that are designed to be activated via normally-closed (n.c.) switches, and a basic circuit of this type is shown in *Figure 2.7*.

In *Figure 2.7*, the coil of a 12V relay is wired in series with the collector of transistor Q1, and bias resistor R1 is wired between the positive supply line and Q1 base. The alarm bell is wired across the supply lines via n.o. relay

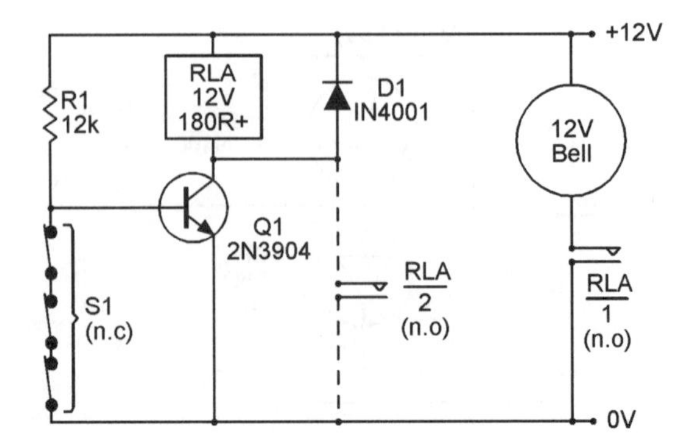

Figure 2.7 *Simple open-to-operate alarm draws a 1mA standby current*

contacts RLA/1, and n.c. operating switch S1 (which may consist of any desired number of n.c. switches wired in series) is wired between the base and emitter of the transistor. Thus, when S1 is closed it shorts the base and emitter of Q1 together, so Q1 is cut off and the relay and the bell are inoperative. Under this condition the circuit draws a quiescent current of 1mA via R1. When S1 opens or a break occurs in its wiring, Q1's base-to-emitter short is removed and the transistor is driven to saturation via R1, thus turning the relay on and activating the alarm bell via relay contacts RLA/1. This basic circuit gives a non-latching type of alarm operation, but can be made to give self-latching operation by wiring a spare set of n.o. relay contacts (RLA/2) between the collector and emitter of Q1, as shown dotted in the diagram.

Thus, the *Figure 2.7* circuit gives fail-safe operation, but draws a quiescent or standby current of 1mA. This standby current can be reduced to a mere 25μA by modifying the circuit in the way shown in *Figure 2.8*. Here, the value of R1 is increased to 470k, and Q1 is used to activate the relay via pnp

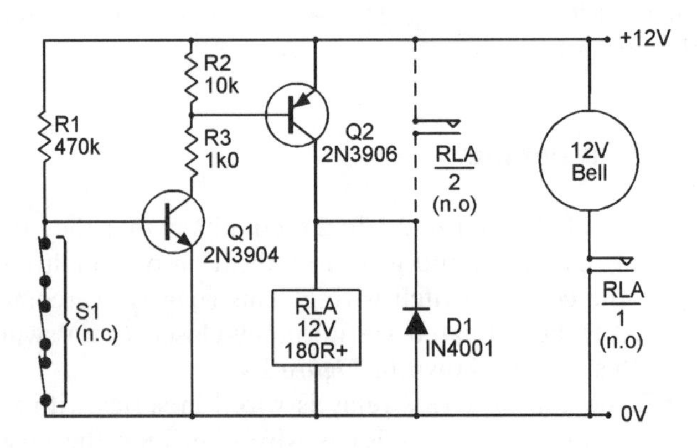

Figure 2.8 *Improved open-to-operate alarm draws a 25μA standby current*

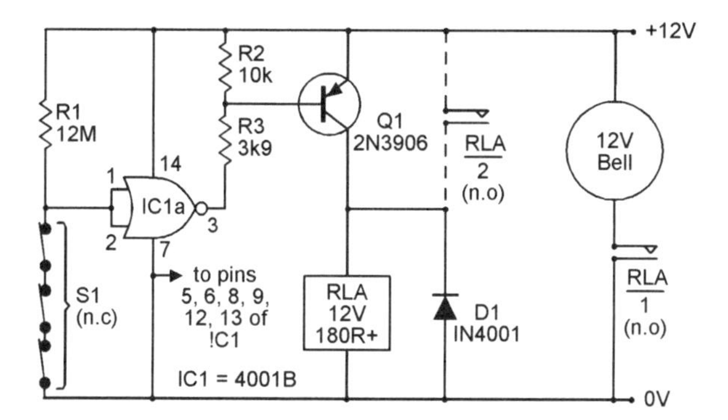

Figure 2.9 *CMOS-aided open-to-operate alarm draws a 1μA standby current*

transistor Q2, and the circuit's action is such that Q1–Q2 and the relay and bell are all off when S1 is closed, but turn on when S1 is open. The basic circuit gives a non-latching form of operation, but can be made self-latching by wiring a spare set of n.o. relay contacts (RLA/2) between the collector and emitter of Q2, as shown dotted in the diagram.

If desired, the standby current of the *Figure 2.8* circuit can be reduced to a mere 1μA or so by using an inverter-connected CMOS gate in place of Q1, as shown in *Figure 2.9*. The gate used here is taken from a 4001B quad 2-input NOR gate IC, and the three unused gates are disabled by shorting their inputs to the 0V line, as shown in the diagram. The used gate has a near-infinite input impedance, and the standby current of the circuit is determined mainly by the R1 value and by the leakage current of Q1. The basic circuit gives a non-latching form of operation, but can be made self-latching by wiring a spare set of n.o. relay contacts (RLA/2) between the collector and emitter of Q1, as shown dotted in the diagram.

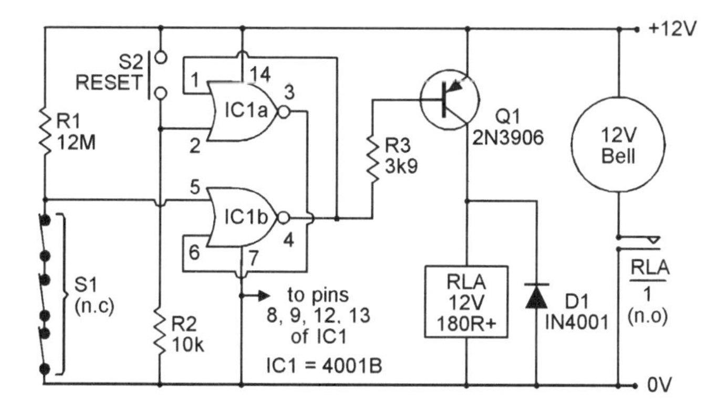

Figure 2.10 *Self-latching CMOS-aided alarm draws a 1μA standby current*

Figure 2.10 shows an alternative way of making the basic *Figure 2.8* circuit give self-latching operation, without resorting to the use of a spare set of n.o. relay contacts. In this case the relay-driving transistor (Q1) is driven by a pair of 4001B CMOS NOR gates that are configured as a bistable multivibrator and has an output that goes low and self-latches if S1 is briefly opened or its leads are broken. As the bistable output goes low it turns Q1 on, thus activating the relay and alarm bell. Once the bistable has latched the bell into the 'on' state, it can be reset into the standby or 'off' mode by closing S1 and momentarily operating RESET switch S2, at which point the bistable's output latches back into the high state and turns off Q1 and the relay and bell. The circuit draws a quiescent current of about 1μA.

'Loop' alarm circuits

One type of contact-operated alarm circuit that is widely used in large shops and stores (and also in domestic garages and garden sheds) is the so-called 'loop' alarm, in which a long length of wire is run out from the alarm unit, is looped through a whole string of 'to be protected' items in such a way that none of them can be removed without cutting or removing the wire, and is then looped back to the alarm unit again, to complete a closed electrical circuit. The alarm sounds instantly if an attempt is made to steal any of the protected items by cutting the wire loop, i.e. by effectively opening its 'contacts'. *Figure 2.11* shows the circuit of a simple battery-powered unit of this type.

The simple *Figure 2.11* loop alarm circuit is a modified version of the self-latching CMOS-aided *Figure 2.9* circuit, with its series-connected S1 security

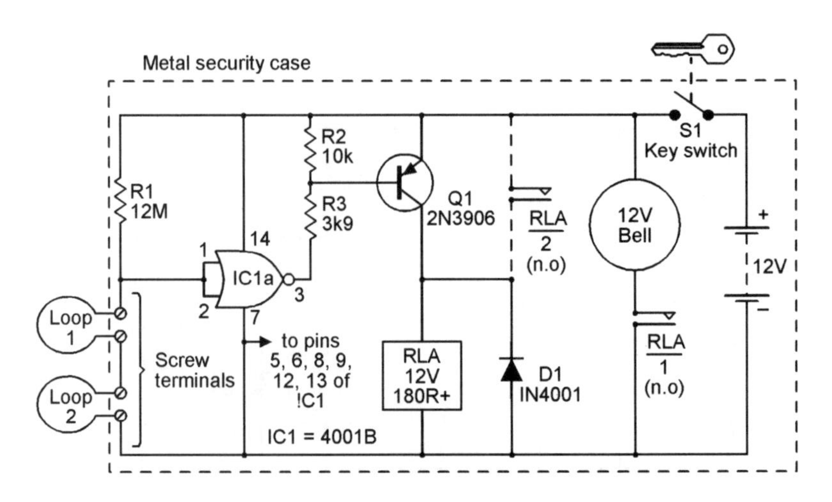

Figure 2.11 *Simple self-latching loop alarm circuit*

switches replaced by a number of series-connected wire 'loops' that – when key-operated switch S1 is closed – activate the self-latching alarm if any part of the loop wiring becomes open circuit. In the diagram, only two loops are shown, but in practice any desired number of loops can be used. The entire circuit (except the loops) is housed inside a metal security case, and the loops are connected to screw terminals on the main circuit board via grommet holes in the side of the case; unwanted loops can be replaced by short circuits connected between the appropriate screw terminals. The entire circuit can be turned on and off via key-switch S1.

Figure 2.12 shows an improved version of the *Figure 2.11* self-latching loop alarm circuit. The first points to note about this version of the circuit are that a LED is connected across the relay coil via R4 and thus illuminates and gives a visual indication whenever the relay is turned on, and that the circuit's +12V power feed is controlled via 4-way key-switch S1 and diodes D2 and D3. When S1 is in position '1', the entire circuit is turned off. When S1 is in position '2', the main part of the circuit (including the LED indicator) is active but the alarm bell and self-latching facility are disabled; this TEST (non-latch) position is meant to be used when testing the loop wiring. When S1 is in the position '3' TEST (latching) position, all of the circuit except the bell is enabled. When S1 is in the position '4' ON position, the entire circuit (including the alarm bell) is enabled, and the circuit gives normal 'security' operation.

The final point to note about the *Figure 2.12* circuit is that n.c. anti-tamper switch S2 is wired in series with the loop network and (when S1 is set to the

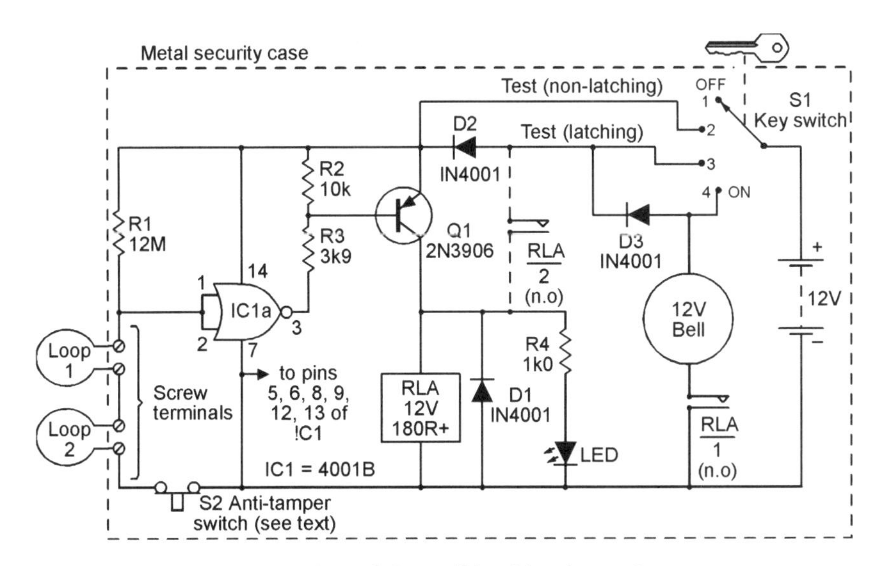

Figure 2.12 *Improved version of the self-latching loop alarm*

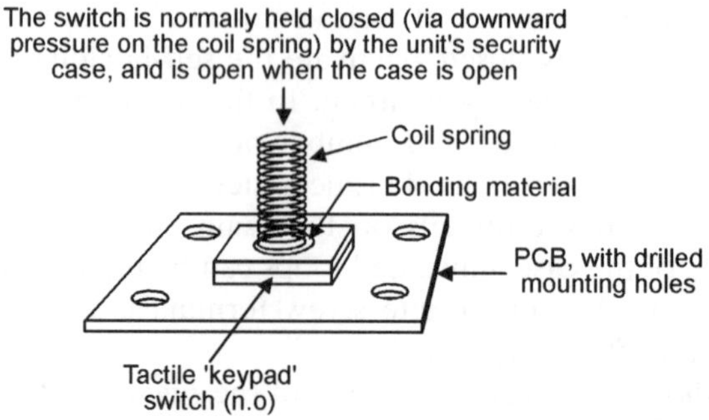

The switch is normally held closed (via downward pressure on the coil spring) by the unit's security case, and is open when the case is open

Coil spring

Bonding material

PCB, with drilled mounting holes

Tactile 'keypad' switch (n.o)

Figure 2.13 *Basic way of constructing an anti-tamper switch (see text)*

ON position) activates the self-latching alarm if it (S2) takes up an 'open' state. S2 is actually an ordinary n.o. tactile 'key-pad' switch with a short coil-spring bonded vertically to its touch-pad, and is fixed to the main circuit board in such a way that the switch is held in the closed n.c. position (via the spring) when the circuit's security case is closed, but opens (thus sounding the alarm) if the case is opened while the alarm system is still turned on. Anti-tamper switches of this basic type are quite easy to make from readily-available components; *Figure 2.13* illustrates the basic method of construction.

Before leaving this Bell and relay-output circuits section of this chapter, note that the various relay-output circuits shown in *Figures 2.2*, *2.3*, and *2.7* to *2.11* can, if desired, be used to activate any type of electrical or electronic alarm or system via their n.o. relay contacts when the relay is triggered in response to an input contact-switching action, and are thus not restricted to use with alarm bells only.

Siren-sound security circuits

Contact-operated security circuits can easily be designed to produce electronically-generated 'siren' alarm sounds in piezoelectric 'sounders' or in electromagnetic loudspeakers. Such systems can be made to produce a

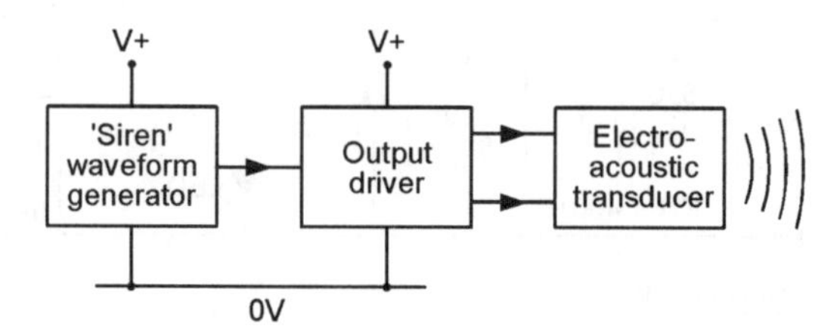

Figure 2.14 *Basic elements of a siren-sound generator*

variety of sounds, at a variety of power levels, and may be designed around various types of semiconductor device. All siren-sound generators take the basic form shown in *Figure 2.14*, and consist of a siren waveform generator, an output driver, and an electro-acoustic transducer.

One of the cheapest and most useful semiconductor devices for use in this type of application is the CMOS 4001B quad 2-input NOR gate IC, which draws near-zero standby current, has an ultra-high input impedance, can operate over a wide range of supply-rail voltages, and can be used in a variety of waveform-generating applications. The rest of this chapter shows various ways of using one or two 4001B ICs and a few other components to make a variety of contact-operated siren-sound security circuits.

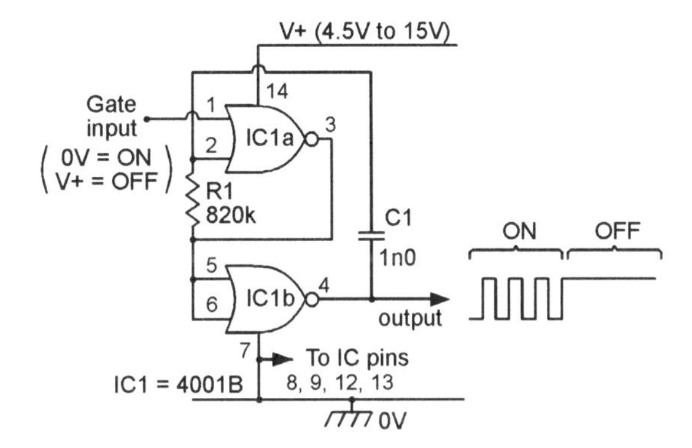

Figure 2.15 *Basic 800Hz monotone 'siren' waveform generator circuit*

Figures 2.15 to *2.17* show three different ways of using 4001B ICs to make practical siren waveform generator circuits. *Figure 2.15* shows the basic circuit of a simple gated 800Hz (monotone) siren waveform generator. Here, two of the gates of a 4001B IC are connected as a gated 800Hz astable multivibrator, and the IC's two remaining gates are disabled by wiring their inputs to ground. The action of this astable is such that it is inoperative, with its pin-4 output terminal locked high (at V+) when its pin-1 input terminal is high (at V+), but acts as a squarewave generator when its input pin is low (at 0V); the generator can thus be gated on and off via the pin-1 input terminal, and produces its output signal on pin-4. The astable's operating frequency is controlled by the R1 and C1 values.

Figure 2.16 shows a single 4001B IC used to make a gated pulsed-tone waveform generator. Here, the two left-hand gates of the IC are wired as a gated low-frequency (about 6Hz) astable squarewave generator, and the two right-hand gates are wired as a gated 800Hz astable that is gated via the 6Hz astable. The action of this circuit is such that it is inoperative, with its pin-11 output terminal locked high (at the positive supply rail voltage) when its

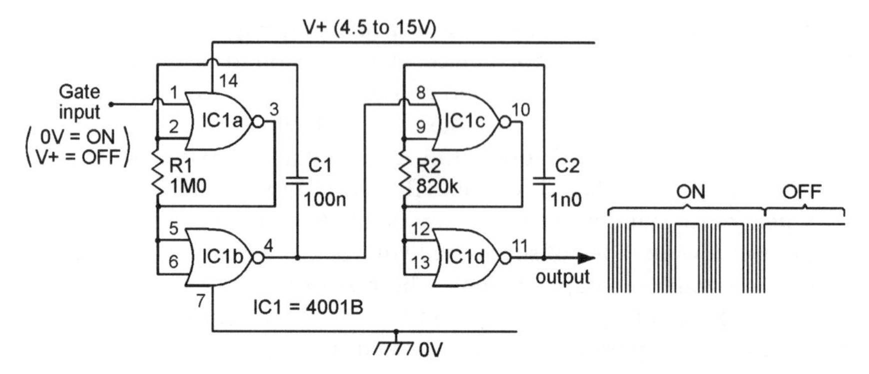

Figure 2.16 *Basic pulsed-tone 'siren' waveform generator circuit*

pin-1 input terminal is high, but becomes active and produces a pulsed-tone output on pin-11 when its input pin is low (at 0V). This generator can thus be gated on and off via the pin-1 input terminal, and when gated on produces an 800Hz tone that is gated on and off at a 6Hz rate. The operating frequency of the 6Hz astable is controlled by R1–C1, and that of the 800Hz astable is controlled by R2–C2.

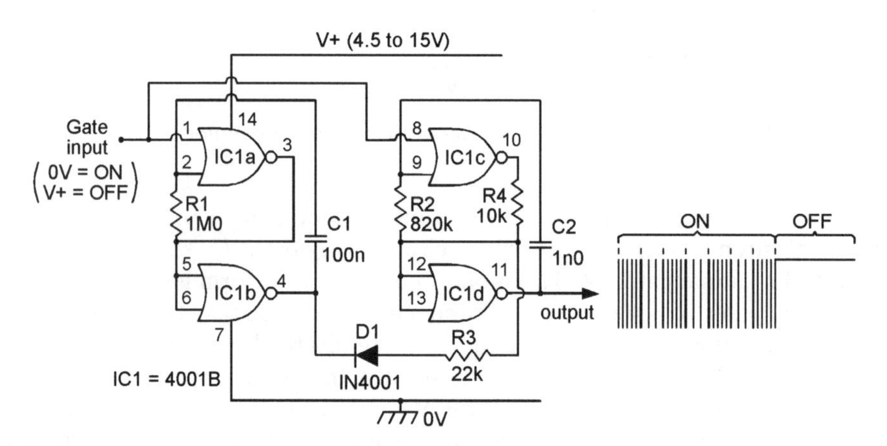

Figure 2.17 *Basic warble-tone 'siren' waveform generator circuit*

Figure 2.17 shows how the *Figure 2.16* circuit can be modified so that it produces a warble-tone alarm signal. These two circuits are basically similar, but in the latter case the 6Hz astable is used to modulate the frequency of the right-hand astable (rather than to simply pulse it on and off), thus causing the generated tone to switch alternately between 600Hz and 450Hz at a 6Hz rate. Note that the pin-1 and pin-8 gate terminals of the two astables are tied together, and both astables are thus activated by the pin-1 'gate' input signal;

the circuit is inoperative, with its pin-11 output terminal locked high (at V+) when the pin-1 input terminal is high, but becomes active and produces a warble-tone output on pin-11 when the input pin is low (at 0V). The operating frequency of this circuit's 6Hz astable is controlled by R1–C1, the centre frequency of the right-hand astable is controlled by R2–C2, and the 'warble-tone' swing of the right-hand astable is controlled via D1–R3.

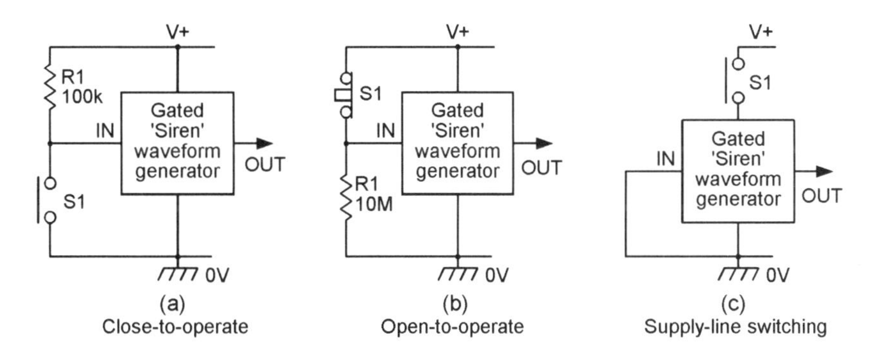

Figure 2.18 *Alternative ways of gating the Figure 2.15 to 2.17 'siren' waveform generator circuits*

Note that each of the *Figure 2.15* to *2.17* gated waveform generator circuits are inactive (with their output terminal locked high) when their pin-1 input terminal is high (at V+), but can be gated on by pulling pin-1 low (to 0V). Each of these circuits can thus be gated on and off by using any of the three input connections shown in *Figure 2.18.* Thus, they can be gated on by closing a n.o. switch by using the input connections shown in *(a)*, or by opening a n.c. switch by using the input connections shown in *(b)*, or can be gated on or off by making or breaking the supply line connection by using the input connections shown in *(c)*. In cases *(a)* and *(b)* the circuit draws a typical standby current of only 1μA or so when in the 'off' state.

If the *Figure 2.15* to *2.17* gated waveform generator circuits are to be used in alarm-sound applications where fairly low acoustic output powers are required, these can be obtained by feeding the circuit's output to a low-cost piezo sounder in any of the three basic ways shown in *Figure 2.19.* Thus, in *(a)* the sounder is driven directly from the generator's output, and in *(b)* it is driven via a 4001B gate that is used as a simple inverting buffer; in both cases the r.m.s. 'alarm' voltage applied across the piezo load equals 50% of the V+ value. In *(c)*, the sounder is driven in the 'bridge' mode via two series-connected 4001B inverters that apply anti-phase signals to the two sides of the piezo load, causing the piezo load to 'see' a squarewave drive voltage with a peak-to-peak value equal to double the V+ value, and an r.m.s. 'alarm' signal voltage that equals the V+ value. The *(c)* circuit thus

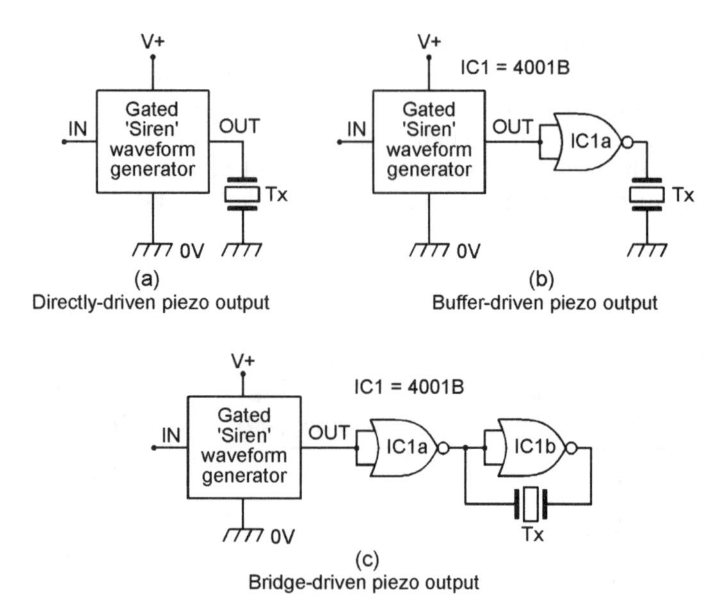

Figure 2.19 *Alternative ways of driving a piezoelectric 'sounder' from the outputs of the Figure 2.15 to 2.17 'siren' waveform generator circuits*

gives four times more acoustic output power than either of the *(a)* or *(b)* circuits.

If the *Figure 2.15* to *2.17* gated waveform generator circuits (which each have an output that is locked high when the generator is gated off) are to be used in alarm-sound applications where fairly high acoustic output powers are required, these can be obtained by feeding the astable's output to inexpensive 'low-fi' or horn-type loudspeakers (these have an electro-acoustic power conversion efficiency that is typically some twenty to forty times greater than a normal hi-fi speaker) via one or other of the simple direct-coupled 'driver' circuits shown in *Figures 2.20* to *2.22*.

Thus, the simple *Figure 2.20* driver circuit is designed to pump a maximum of only a few hundred milliwatts of audio power into a cheap 64R speaker. When the siren waveform generator is gated off its output is high and Q1 is thus cut off, but when the generator is gated on its output drives Q1 on and off and causes it to feed power to the 64R speaker. The output power depends on the supply rail voltage, and has a value of about 520mW at 12V, or 120mW at 6V, when feeding a 64R speaker load. Note that, since Q1 is used as a simple power switch in this application, very little power is lost across the 2N3906 transistor, but its current rating (200mA maximum) may be exceeded if the circuit is used with a supply value greater than 12V.

The *Figure 2.21* driver circuit can pump a maximum of 6.6 watts of audio power into an 8R0 speaker load, or 3.3 watts into a 16R load. Here, both

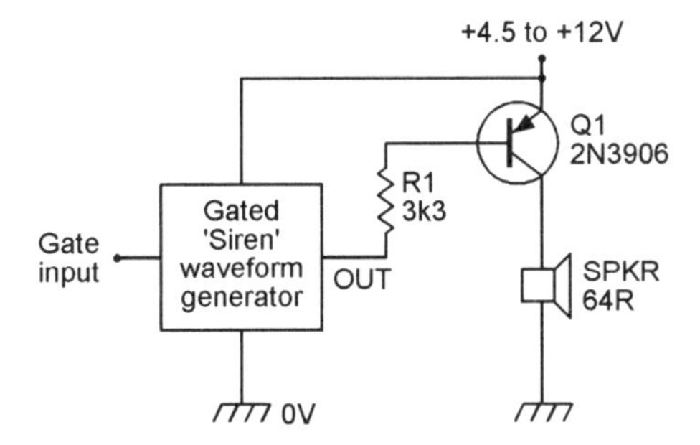

Figure 2.20 *Simple output driver circuit that can feed up to 520mW into a 64R speaker load*

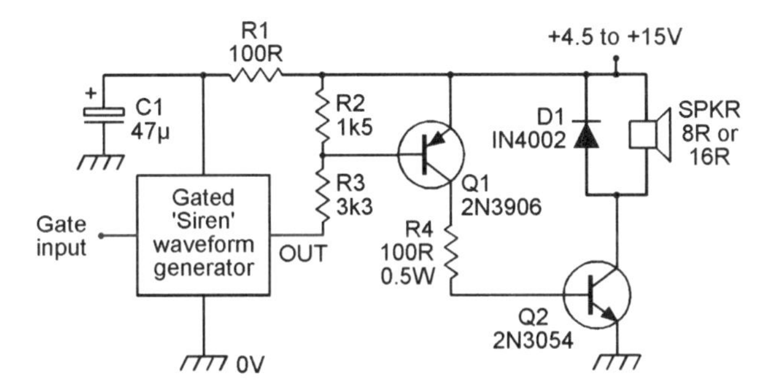

Figure 2.21 *Medium power (up to 6.6 watts into 8R0) output driver*

transistors are cut off when the waveform generator is gated off, but are switched on and off in sympathy with the siren waveform when the generator is gated on. Note in this circuit that the positive power supply rail is fed directly to the output driver, but is fed to the waveform generator via decoupling network R1–C1, that voltage divider R2–R3 ensures that the output stages are not driven on until the generator's output voltage falls at least 1.9V below thc supply rail value, and that diode D1 is used to damp the speaker's back-emf when driver Q2 switches off.

Finally, the *Figure 2.22* driver circuit can pump a maximum of 13.2 watts into a 4R0 speaker load when powered from a 15V supply. Here, all three transistors are cut off when the waveform generator is cut off, but are switched on and off in sympathy with the siren waveform when the generator is gated on.

Thus, *Figures 2.15* to *2.17* show three alternative 'siren' waveform generator circuits that can – when used in practical contact operated security circuits – each be gated in any of three basic ways and be used in conjunction with any of six basic types of acoustic output driver circuit, thus offering a total of 54 different circuit combinations. *Figure 2.23*, for example,

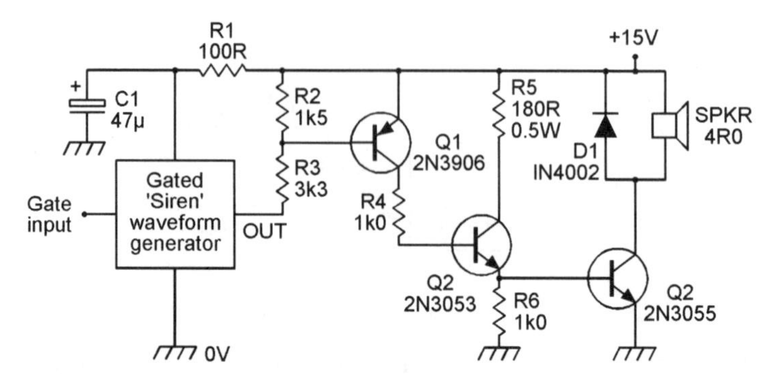

Figure 2.22 *High power (up to 13.2 watts into 4R0) output driver*

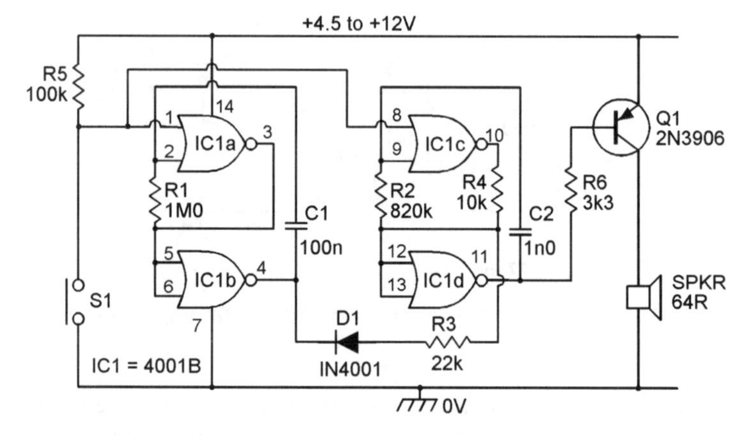

Figure 2.23 *Low-power (up to 520mW) warble-tone alarm-call generator, activated by closing a n.o switch*

shows how the *Figure 2.17*, *2.18(a)* and *2.20* circuits can be combined to make a warble-tone alarm-call generator that can be activated by closing a n.o. switch and which can pump 520mW into a 64R speaker load when operated from a 12V supply.

Self-latching siren-sound generator circuits

The siren-sound generator circuits shown in *Figures 2.15* to *2.17* are all non-latching types that produce an output only while activated by their control switches. By contrast, *Figures 2.24* and *2.25* show circuits that give some form of self-latching siren-sound waveform generating action.

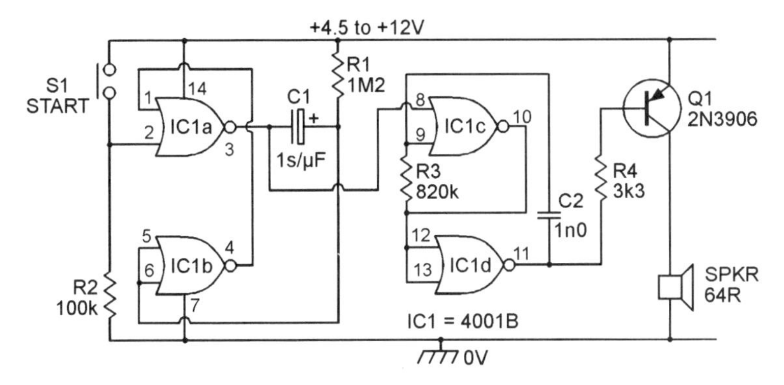

Figure 2.24 *Low-power one-shot 800Hz monotone alarm-call generator*

The *Figure 2.24* circuit is that of a one-shot or auto-turn-off monotone siren waveform generator. Here, IC1a–IC1b are wired as a one-shot (monostable) multivibrator that can be triggered by a rising voltage on pin-2, and IC1c–IC1d are wired as a gated 800Hz astable multivibrator that is activated by the output of the monostable. Thus, the circuit action is such that both multivibrators are normally inoperative and the circuit consumes only a small leakage current. As soon as S1 is momentarily closed, however, the monostable triggers and turns on the 800Hz astable, which then continues to operate for a pre-set period, irrespective of the state of S1. At the end of this period the alarm signal automatically turns off again, and the action is complete. The circuit can be retriggered again by applying another rising voltage to pin-2 via S1. The alarm duration time is determined by C1, and approximates one second per microfarad of C1 value; periods of several minutes can readily be obtained.

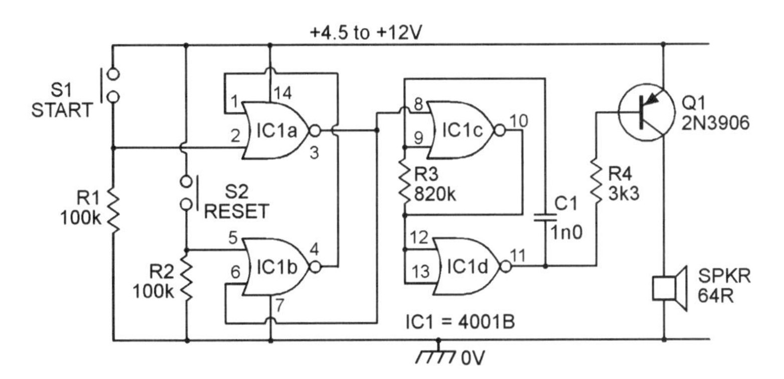

Figure 2.25 *Low-power self-latching 800Hz monotone alarm-call generator*

Figure 2.25 shows the circuit of a true self-latching 800Hz switch-activated monotone siren waveform generator. Here, IC1a–IC1b are wired as a manually-triggered bistable multivibrator, and IC1c–IC1d as a gated 800Hz astable that is activated via the bistable. The circuit action is such that the bistable output is normally high and the astable is disabled, and the circuit consumes only a small leakage current. When S1 is briefly operated, pin-2 of the IC is pulled high, so the bistable changes state and its output locks low and activates the 800Hz astable. Once the alarm signal has been so activated, it can only be turned off again by removing the positive signal from pin-2 and briefly closing RESET switch S2, at which point the circuit resets and its current consumption returns to leakage levels.

Note that the outputs of the above two self-latching waveform generator circuits can be converted into acoustic form via any of the basic driver circuits shown in *Figures 2.19* to *2.22*.

Multitone alarms

Each of the *Figure 2.15* to *2.17* and *Figure 2.24–2.25* circuits has a single 'start' switch and – when connected to a suitable driver and transducer – generates a unique sound when that switch is operated. By contrast, *Figures 2.26* and *2.27* show a couple of 'multitone' alarm-call generators that each have two or three input switches and which generate a unique sound via each input switch. These circuits are useful in identifier applications, such as in door announcing where, for example, a high tone may be generated via a front door switch, a low tone via a back door switch, and a medium tone via a side door switch.

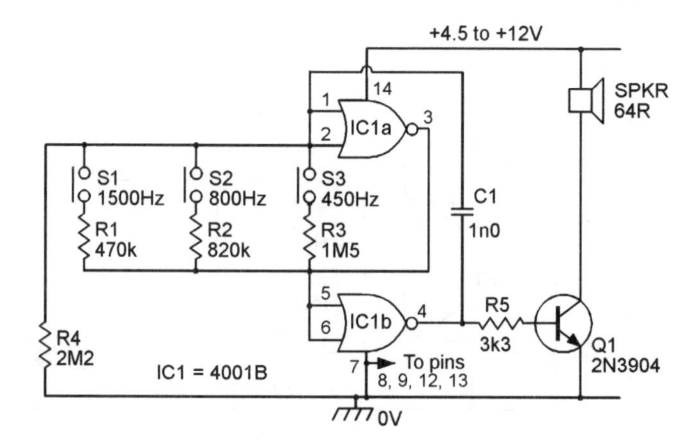

Figure 2.26 *Low-power 3-input multitone alarm-call generator*

The *Figure 2.26* circuit is that of a simple 3-input monotone alarm-call generator. Here, two 4001B gates are wired as a modified astable multivibrator, and the action is such that the circuit is normally inoperative and drawing only a slight standby current, but becomes active and acts as a squarewave generator when a resistance is connected between pins 2 and 5 of the IC. This resistance must be less than the 2M2 value of R4, and the frequency of the generated tone is inversely proportional to the resistance value used. With the component values shown, the circuit generates a tone of about 1500Hz via S1, 800Hz via S2, and 450Hz via S3. These tones are each separated by about an octave, so each push-button generates a very distinctive tone. The basic circuit generates an output power of about 520mW into a 64R speaker when powered from a 12V supply, and this is adequate for most practical applications.

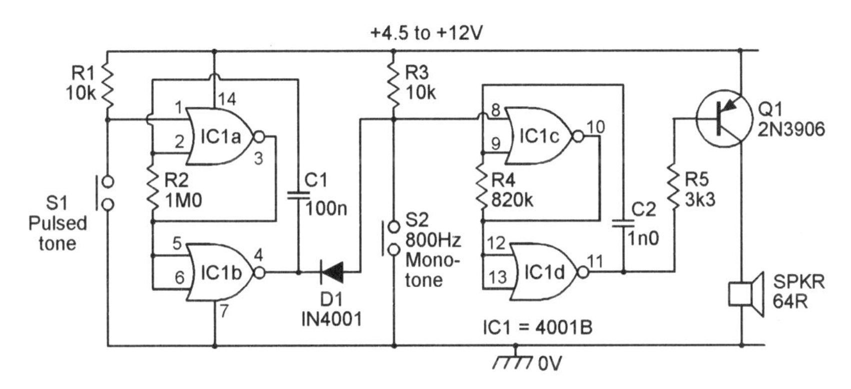

Figure 2.27 *2-input multitone alarm-call generator*

Figure 2.27 shows a 2-input multitone unit that generates a pulsed-tone signal via S1 or a monotone signal via S2. Here, IC1a–IC1b are wired as a gated 6Hz astable and IC1c–IC1d as a 800Hz astable. The astables are interconnected via D1, and the circuit action is such that the 6Hz astable activates the 800Hz astable when S1 is closed, thus producing a pulsed-tone output signal, but only the 800Hz astable operates when S2 is closed, thus producing a monotone output signal. This basic circuit generates an output power of about 520mW into a 64R speaker when powered from a 12V supply, which is adequate for most practical applications, but the power level can, if desired, be greatly boosted via the circuits of *Figures 2.21* or *2.22*.

3

Optoelectronic security circuits

Optoelectronic circuits – which respond to either visible or invisible (usually infra-red) light levels – are widely used in modern domestic and commercial security systems. Optoelectronic security circuits come in three basic types, the first of which is the simple 'visible-light-level' type that reacts when a light level goes above or below a preset value. The second type is the light-beam alarm, which reacts when a person, object or animal breaks or reflects a projected visible or infra-red (IR) light beam. The third type is the passive infra-red (PIR) detector, which is sensitive to the heat-generated infra-red energy radiated by the human body, and thus reacts when a human or other large warm-blooded animal comes within the sensing range of the PIR detector. Practical optoelectronic circuits of all of these types are described in this chapter.

'Visible-light-level' circuits

LDR basics

Most optoelectronic circuits that are designed to respond to normal visible light use a cadmium sulphide (CdS) LDR (light-dependent resistor) as their light-sensing element or photocell. *Figure 3.1* shows an LDR's circuit symbol

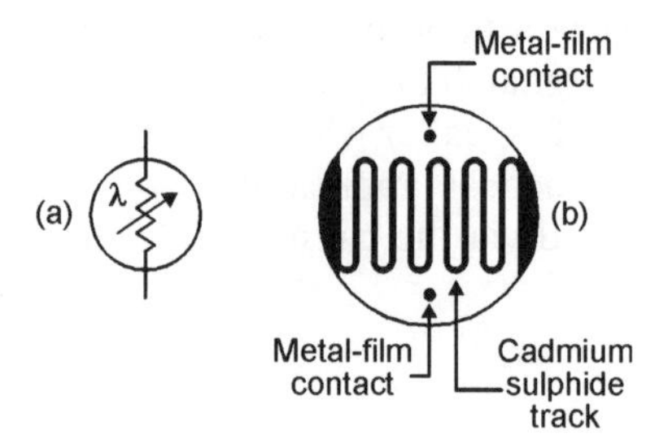

Figure 3.1 *LDR symbol (a) and basic structure (b)*

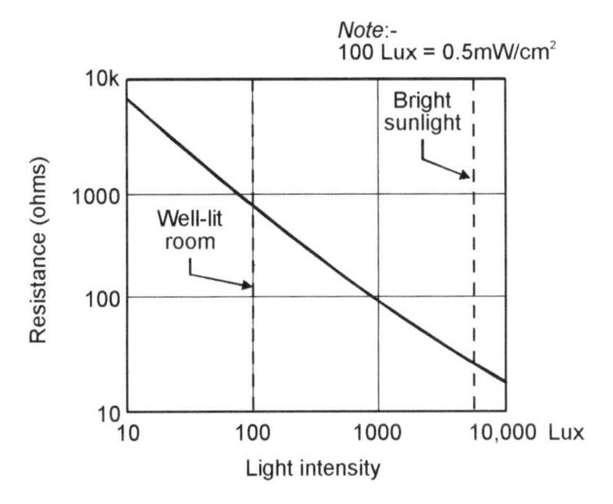

Figure 3.2 *Typical characteristics curve of an LDR with a 10mm face diameter*

and basic construction, which consists of a pair of metal film contacts separated by a snake-like track of light-sensitive cadmium sulphide film; the structure is housed in a clear plastic or resin case. *Figure 3.2* shows the typical photoresistive graph that applies to an LDR with a face diameter of about 10mm; the LDR acts as a high resistance (typically hundreds of kilohms) under dark conditions, falling to about 900R at a light intensity of 100 Lux (typical of a well lit room) or about 30R at 8000 Lux (typical of bright sunlight). Note that although most of the practical circuits shown throughout this 'Visible-light-level' section of this chapter are shown using an ORP12 photocell, they will in fact work well with almost any general-purpose LDRs with face diameters in the range 3mm to 12mm.

Simple relay switches and alarms

The simplest 'visible-light-level' security circuits are the types that are designed to activate when the light level goes through a large change, such as from near-dark to a moderately high level, or vice versa. Some circuits of this type function as 'light-activated' units that switch on a relay or safety mechanism or sound an alarm when light enters a normally dark area such as the inside of a storeroom, wall safe, or security case, etc. Other circuits are designed as 'dark-activated' units, and activate when the light level drops well below some nominal value. *Figures 3.3* to *3.7* show some simple circuits of these types.

The *Figure 3.3* circuit is that of a simple non-latching light-activated relay switch. Here, R1–LDR and R2 form a potential divider that controls the base-bias of Q1. Under dark conditions the LDR resistance is very high, so negligible base-bias is applied to Q1, and Q1 and relay RLA are both off.

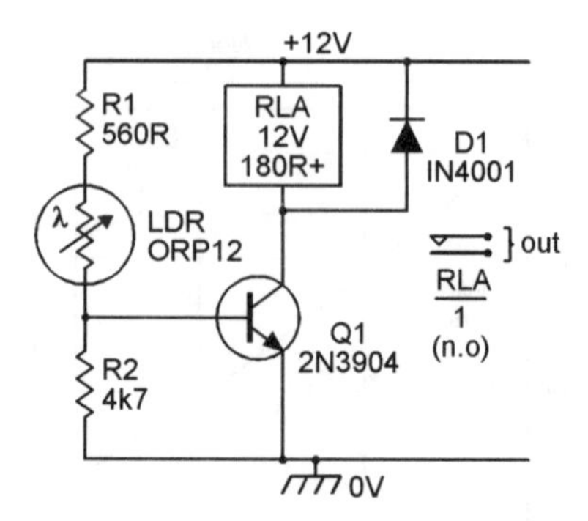

Figure 3.3 *Simple non-latching light-activated relay switch*

When light enters the normally dark area and falls on the LDR face, however, the LDR resistance falls to a fairly low value and base-bias is applied to Q1, which thus turns on and activates the relay and its RLA/1 contacts, which can be used to control external circuitry. The relay can be any 12V type with a coil resistance of 180R or greater.

The simple *Figure 3.3* circuit has a fairly low sensitivity, and has no facility for sensitivity adjustment. *Figure 3.4* shows how these deficiencies can be overcome by using a super-alpha-connected pair of transistors in place of Q1, and by using sensitivity control RV1 in place of R2; this circuit can be activated by LDR resistances as high as 200k (i.e. by exposing the LDR to very small light levels), and draws a standby current of only a few microamps under 'dark' conditions. The diagram also shows how the circuit can be made to give a self-latching action via relay contacts RLA/2; normally-closed push-button switch S1 enables the circuit to be reset (unlatched) when required.

Figure 3.5 shows an LDR used in a simple dark-activated circuit that turns relay RLA on when the light level falls below a value that is pre-set via RV1.

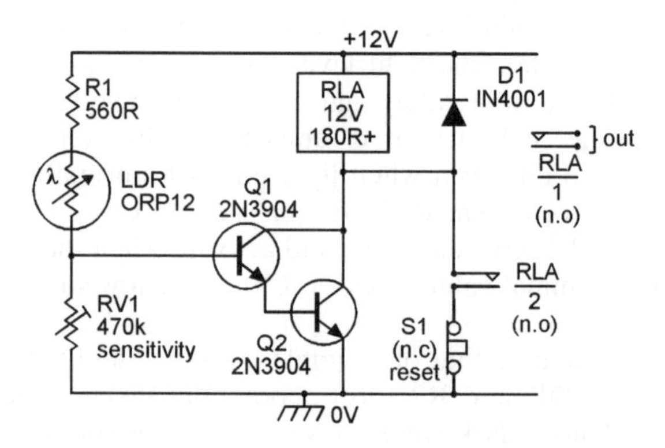

Figure 3.4 *Sensitive self-latching light-activated relay switch*

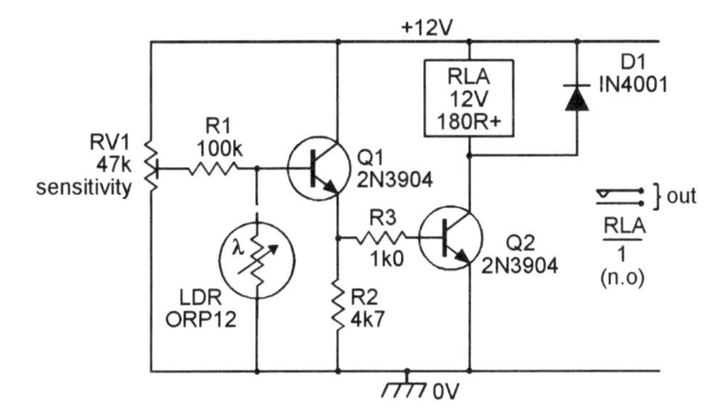

Figure 3.5 *Simple dark-activated relay switch*

Here, R1 and LDR act as a potential divider that is fed with a pre-set dc voltage from RV1 and generates an output voltage that rises as the light level falls, and vice versa. This generated voltage is buffered by emitter follower Q1 and used to control the relay via common-emitter amplifier Q2 and current-limiting resistor R3. Thus, when the light level falls below the pre-set value the R1–LDR divider's output voltage reaches a large enough level to activate both Q1 and Q2, and the relay turns on.

Figures 3.6 and *3.7* show simple SCR (silicon controlled rectifier) circuits that – under the light-sensitive 'alarm' condition – activate an ordinary self-interrupting electromechanical alarm bell. This bell must (when the specified C106D SCR is used) consume an operating current of less than 2A, and the circuit's supply voltage must be at least 1V greater than the bell's nominal operating voltage.

The *Figure 3.6* circuit is a simple non-latching one, in which the alarm bell is wired in series with an SCR that receives its gate drive from the positive supply rail via the R1–LDR–R2 potential divider. Thus, under dark conditions

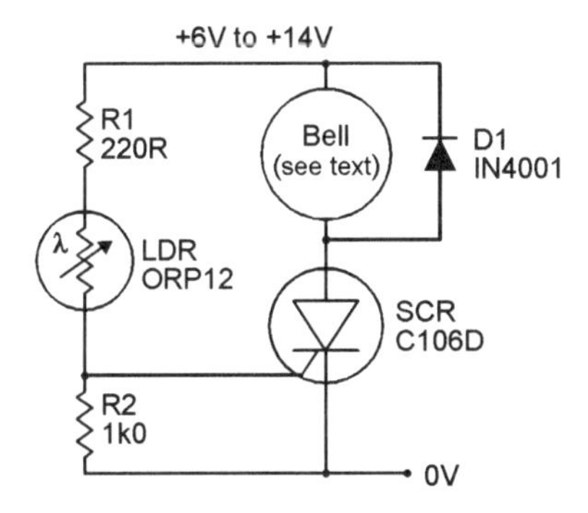

Figure 3.6 *Simple light-activated alarm bell*

the LDR resistance is very high, so negligible gating voltage appears across R2, and the SCR and alarm are both off. When light enters the normally dark area and falls on the LDR face, however, the LDR resistance falls to a fairly low value, and if this value is below roughly 10k the LDR passes enough current to gate the SCR into the 'on' mode, thus sounding the alarm bell. Most LDRs (including the ORP12) give a resistance of less than 10k when exposed to low-intensity room lighting or the light of a torch, so this circuit operates as soon as it is exposed to a moderate amount of illumination.

Note in the above circuit that, although the SCR is a self-latching device, it automatically unlatches each time that the bell enters a self-interrupt phase (and the SCR anode current drops to zero). Consequently, the alarm bell automatically turns off when the light level falls back below the SCR's 'trip' level.

The *Figure 3.6* circuit has a fairly low sensitivity and has no facility for sensitivity adjustment. *Figure 3.7* shows how these deficiencies can be overcome by using RV1 in place of R2 and by using Q1 as a buffer between the LDR and the SCR gate; this circuit can be activated by LDR resistances as high as 200k (i.e. by exposing the LDR to very small light levels). This diagram also shows how the circuit can be made self-latching by wiring R4 across the bell so that the SCR anode current does not fall to zero as the bell self-interrupts. Switch S1 enables the circuit to be reset (unlatched) when required.

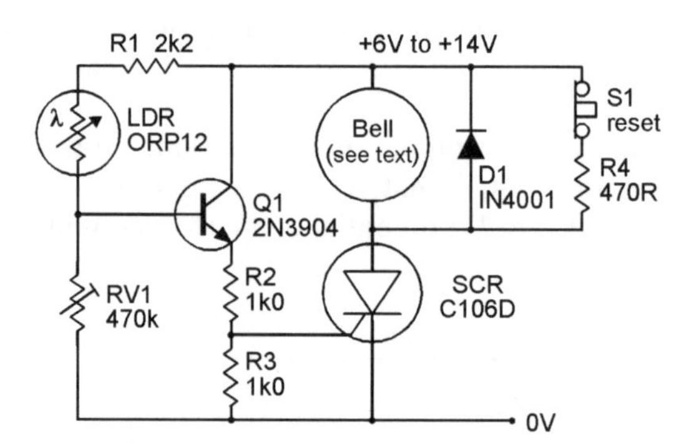

Figure 3.7 *Improved light-activated alarm bell with self-latching facility*

Siren-output alarms

The *Figure 3.2* to *3.7* circuits are designed to give either alarm-bell or relay outputs. In some applications an electronic 'siren sound' output may be preferable, and *Figures 3.8* and *3.9* show low-power (up to 520mW) speaker-

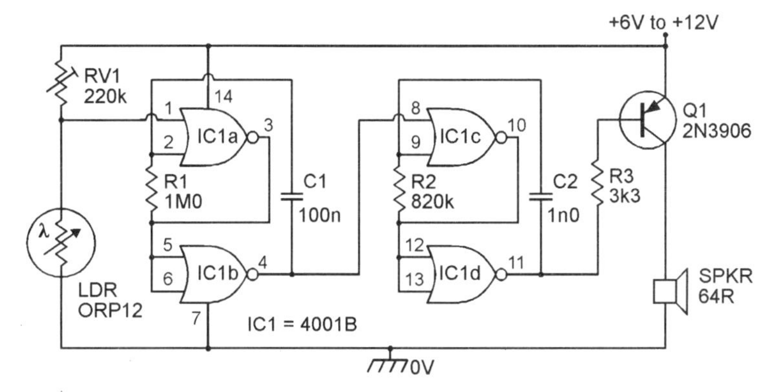

Figure 3.8 *Non-latching light-activated alarm with pulsed-tone output*

output circuits of this type. The *Figure 3.8* circuit gives non-latching operation and develops a pulsed-tone output signal. The *Figure 3.9* circuit gives self-latching operation, and develops a monotone output signal. Both circuits are designed around a 4001B CMOS IC.

In the *Figure 3.8* circuit the IC's two left-hand gates are wired as a low-frequency (6Hz) gated astable multivibrator that is activated via the light-sensitive LDR–RV1 potential divider, and the two right-hand gates are wired as a gated 800Hz astable that is activated by the 6Hz astable. Under dark conditions the LDR–RV1 output voltage is high and both astables are gated off, and the circuit consumes a fairly low standby current, but under bright conditions the LDR–RV1 output voltage is low and both astables are activated, with the left-hand astable gating the 800Hz astable on and off at a 6Hz rate, thus generating a pulsed 800Hz tone in the speaker.

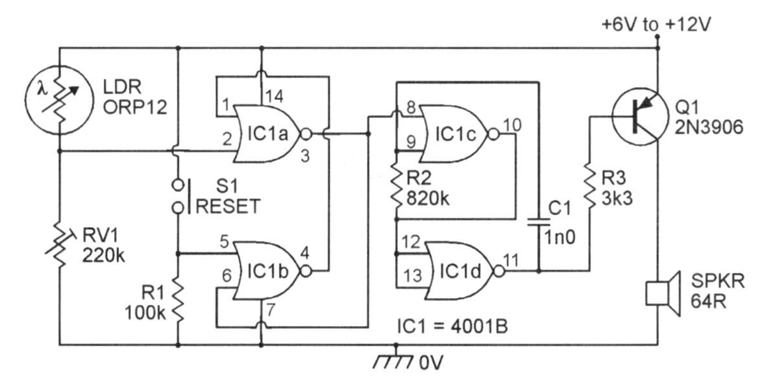

Figure 3.9 *Self-latching light-activated alarm gives monotone output*

In the self-latching *Figure 3.9* circuit the IC's two left-hand gates are wired as a simple bistable multivibrator that can be activated (SET) by a high voltage on the output of the light-sensitive LDR–RV1 potential divider, and the two right-hand gates are wired as a gated 800Hz astable that is activated by the bistable's output. Under normal dark conditions the output of the LDR–RV1 divider is low, the bistable output is in the high 'RESET' state, thus gating the astable off, and the circuit consumes only a small standby current. When the LDR is exposed to light, however, the LDR–RV1 junction goes high, flipping the self-latching bistable into the SET mode, in which its output goes low and gates on the 800Hz astable, thus generating a monotone speaker signal. Once it has been activated, the circuit can only be turned off again by removing the LDR's illumination and briefly closing RESET switch S1, at which point the bistable output resets to the high state and the astable is gated off again.

Note that the trigger points of the *Figure 3.8* and *3.9* circuits occur at the point at which the LDR–RV1 junction voltage swings above or below IC1's 'transition' voltage value, which lies somewhere between one-third and two-thirds of the supply voltage value. In practice, the light-trigger points of these circuits are not greatly affected by supply voltage variations. Also note that, when using the output stages shown in the diagrams, these two circuits each give maximum output powers of only 520mW into the 64R speakers, but that these powers can be boosted to as high as 13.2 watts by replacing the Q1 output stages with the power-boosting circuits shown in *Figures 2.21* or *2.22* (in Chapter 2).

Precision relay-switching circuits

The relay-switching circuits of *Figures 3.3* to *3.5* are fairly sensitive to variations in supply voltage and temperature, and are not suitable for use in a *precision* light-sensing application. *Figure 3.10* shows a very sensitive light-

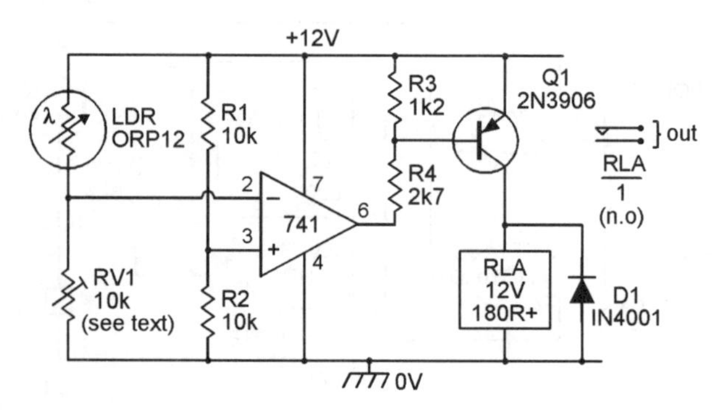

Figure 3.10 *Precision light-activated relay switch*

sensitive relay-driving circuit that is not influenced by such variations, and is thus suitable for use in many precision light-sensing applications, and which turns the relay on when the light level rises above a pre-set value. In this case, LDR–RV1 and R1–R2 are connected in the form of a Wheatstone bridge, and the op-amp and Q1–RLA act as a highly sensitive balance-detecting switch. The bridge balance point is quite independent of variations in supply voltage and temperature, and is influenced only by variations in the relative values of the bridge components.

In *Figure 3.10*, the LDR and RV1 form one arm of the bridge, and R1–R2 form the other arm. These arms can actually be regarded as potential dividers, with the R1–R2 arm applying a fixed half-supply voltage to the non-inverting input of the op-amp, and with the LDR–RV1 divider applying a light-dependent variable voltage to the inverting terminal of the op-amp. In use, RV1 is adjusted so that the LDR–RV1 voltage rises slightly above that of R1–R2 as the light intensity rises to the desired trigger level, and under this condition the op-amp output switches to negative saturation and thus drives RLA on via Q1 and biasing resistors R3-R4. When the light intensity falls below this level, the op-amp output switches to positive saturation, and under this condition Q1 and the relay are off.

The *Figure 3.10* circuit is very sensitive and can detect light-level changes too small to be seen by the human eye. The circuit can be modified to act as a precision dark-activated switch (which turns the relay on when the light level falls below a pre-set value) by either transposing the inverting and non-inverting input terminal connections of the op-amp, or by transposing RV1 and the LDR. *Figure 3.11* shows a circuit using the latter option. This diagram also shows how a small amount of hysteresis can be added to the circuit via feedback resistor R5, so that the relay turns on when the light level falls to a particular value, but does not turn off again until the light level rises a substantial amount above this value (this action helps prevent spurious switching due to passing shadows, etc., in automatic lighting-control

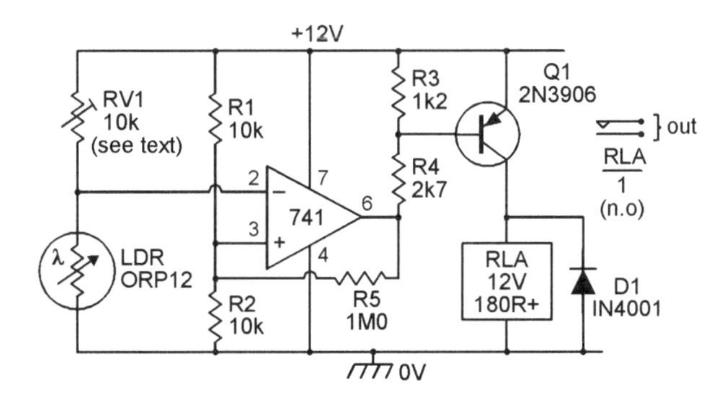

Figure 3.11 *Precision dark-activated switch, with hysteresis*

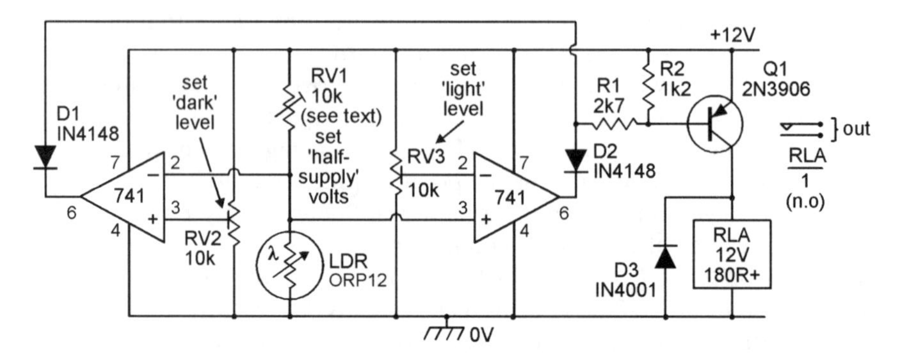

Figure 3.12 *Combined light-/dark-activated switch with a single relay output*

circuits). The hysteresis magnitude is inversely proportional to the R5 value, being zero when R5 is open circuit.

Figure 3.12 shows how a precision combined light-/dark-activated switch (which activates a single relay if the light intensity rises above one pre-set value or falls below another pre-set value) can be made by combining the 'light' and 'dark' switch circuits of *Figures 3.10* and *3.11*. To set up this circuit, first pre-set RV1 so that roughly half-supply volts appear on the LDR–RV1 junction when the LDR is illuminated at the mean or normal intensity level. RV2 can then be pre-set so that RLA turns on when the light intensity falls to the desired dark level, and RV3 can be pre-set so that RLA activates when the light rises to the desired 'bright' level.

Note in the *Figure 3.10* to *3.12* 'precision' circuits that the basic RV1 value should be roughly double that of the LDR resistance value at the desired 'trigger' light levels, so that RV1's slider is reasonably close to its central position under this condition. Thus, the RV1 value may have to be reduced below 10k if a circuit is set to trigger under very bright conditions, or may have to be increased above 10k if used to trigger under very dark conditions.

Precision alarm-sounding circuits

The basic op-amp based precision circuits of *Figures 3.10* to *3.12* can easily be used in conjunction with some of the alarm-bell activator or siren-sound generator circuits already shown in this chapter, to make various precision alarm-sounding circuits. *Figure 3.13*, for example, shows modified versions of the basic *Figure 3.6* and *3.10* circuits combined to make a precision light-activated alarm bell circuit that can easily be changed into a dark-activated unit by simply transposing RV1 and the LDR. Note that the op-amp's supply line is decoupled from that of the bell via D3 and C1.

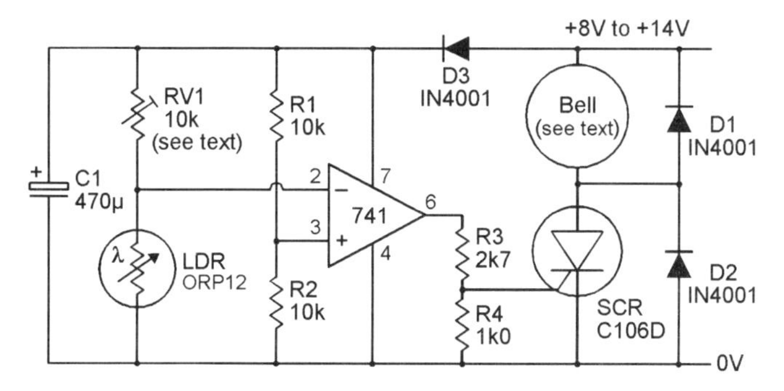

Figure 3.13 *Precision light-activated alarm bell*

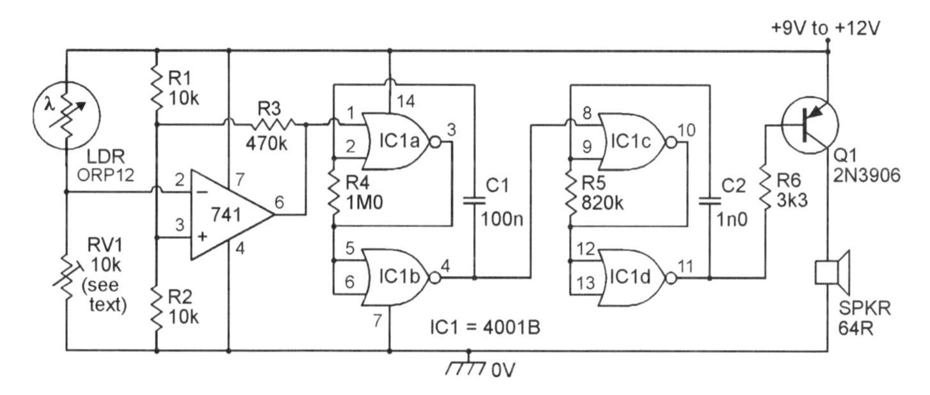

Figure 3.14 *Precision light-activated pulsed-tone alarm with hysteresis*

Similarly, *Figure 3.14* shows modified versions of the basic *Figure 3.8* and *3.10* circuits combined to make a precision light-activated pulsed-tone alarm with built-in hysteresis. The hysteresis is controlled by R3, which can have its value altered to obtain different hysteresis values, or can be removed if hysteresis is not needed.

Light-beam alarm circuits

Simple visible-light designs

A light-beam alarm system consists of a focused light-beam transmitter (Tx) and a focused light-beam receiver (Rx), and may be configured to give either a direct-light-beam or a reflected-light-beam type of optical contact

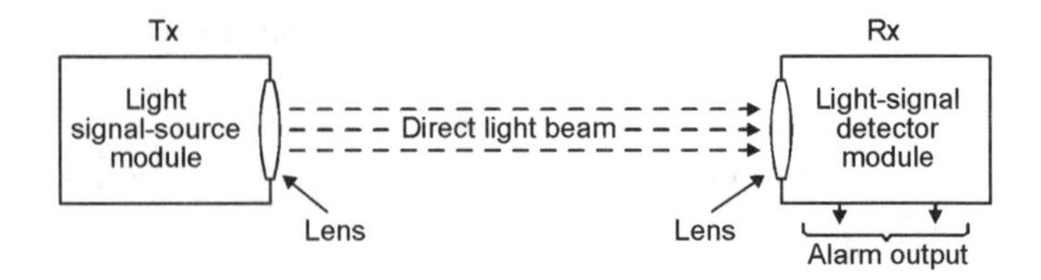

Figure 3.15 *The basic elements of a direct-light-beam type of alarm system*

operation. *Figure 3.15* shows the basic elements of a direct-light-beam type of alarm system, in which the sharply focused Tx light-beam is aimed directly at the light-sensitive input point of the Rx unit, which (usually) is designed to activate an external alarm or safety mechanism if a person, object, or piece of machinery enters the light-beam and breaks the optical contact between the Tx and Rx.

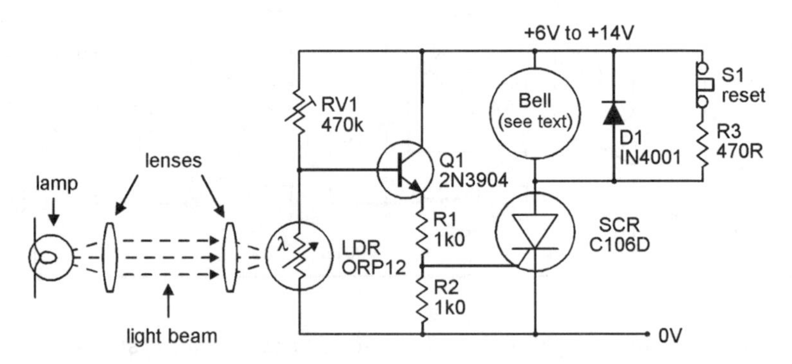

Figure 3.16 *Simple direct-light-beam alarm with alarm bell output*

Figure 3.16 shows a very simple example of a lamp-and-LDR direct-light-beam system that activates an alarm-bell if the beam is interrupted. The beam is generated via an ordinary electric lamp and a lens, and is focused (via a 'collector' lens) on to the face of an LDR in the remote Rx unit, which operates as a dark-activated alarm. Normally, the LDR face is illuminated by the light-beam, so the LDR has a low resistance and very little voltage thus appears on the RV1–LDR junction, so the SCR and bell are off. When the light-beam is broken, however, the LDR resistance goes high and enough voltage appears on the RV1–LDR junction to trigger the SCR, which drives the alarm bell on; R3 is used to self-latch the alarm.

Figure 3.17 shows the basic elements of a reflected-light-beam type of alarm system, in which the Tx light-beam and Rx lens are optically screened from each other but are both aimed outwards towards a specific point, so that an optical link can be set up by a reflective object (such as metallic paint

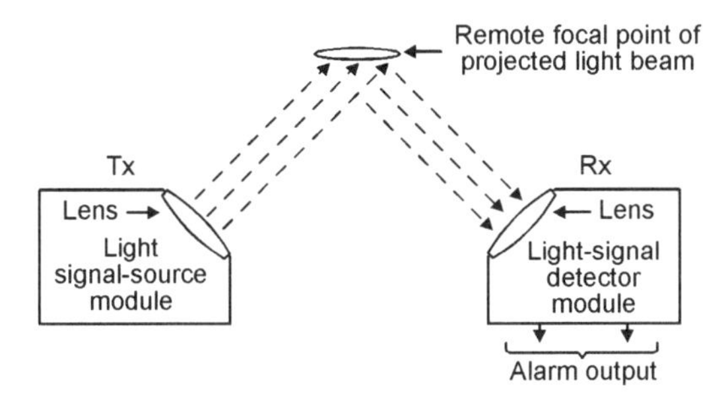

Figure 3.17 *The basic elements of a reflected-light-beam type of alarm system*

or smoke or fog particles) placed at that point. This type of system is usually designed to activate an alarm when the presence of such an object is detected, but can also be configured to give the reverse action, so that the alarm activates if a reflective object is legally or illegally removed.

Units of the *Figure 3.17* type were once widely used in conjunction with reflection-type fog and smoke detector units; *Figure 3.18* shows a sectional view of a smoke detector unit of this type. Here, the lamp and LDR are mounted in an open-ended but light-excluding box, in which an internal screen prevents the lamp-light from falling directly on the LDR face. The lamp is a source of both light and heat, and the heat causes convection currents of air to be drawn in from the bottom of the box and to be expelled through the top. The inside of the box is painted matt black, and the construction lets air pass through the box but excludes external light.

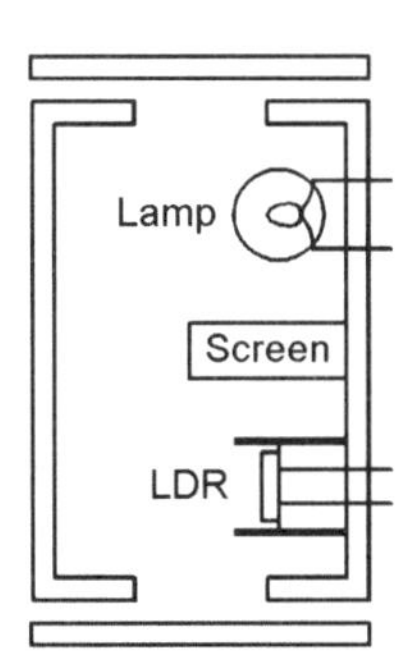

Figure 3.18 *Sectional view of a reflection-type smoke detector*

Thus, if the convected air currents are smoke-free, no light falls on the LDR face, and the LDR presents a high resistance. If the air currents do contain smoke, however, the smoke particles cause the light of the lamp to reflect on to the LDR face and so cause a large and easily detectable

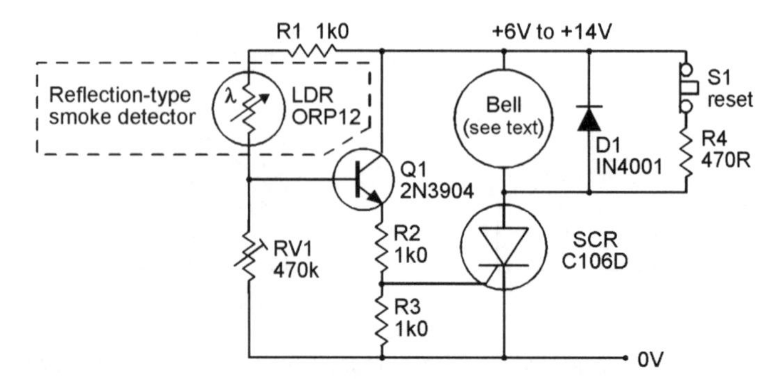

Figure 3.19 *Smoke alarm with alarm bell output*

decrease in the LDR resistance. *Figure 3.19* shows the practical circuit of a reflection-type smoke alarm that can be used with this detector; the circuit acts in the same way as the improved *Figure 3.7* light-activated alarm circuit.

IR light-beam alarm basics.

Simple lamp-and-LDR light-beam alarms of the types described in the last section have several obvious disadvantages in most modern security-alarm applications. Their light-beams are, for example, clearly visible to an intruder, the transmitter's filament lamp is unreliable, and the systems are (because the transmitter's filament lamp wastes a lot of power) very inefficient. In practice, virtually all modern light-beam security systems operate in

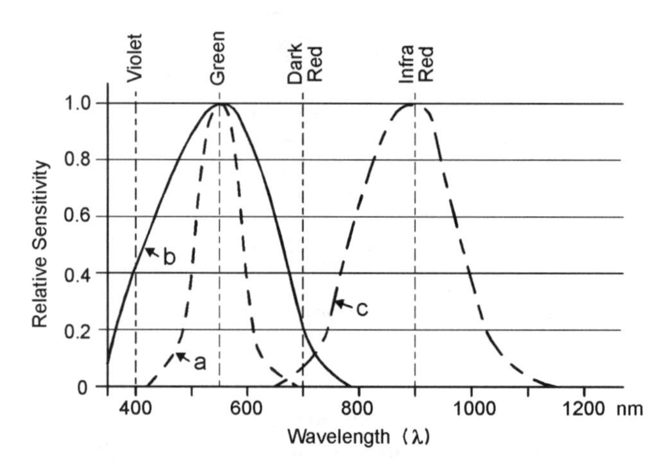

Figure 3.20 *Typical spectral response curves of (a) the human eye and (b) general-purpose and (c) IR photodiodes/phototransistors*

the invisible infra-red (rather than visible light) range, and use one or more pulse-driven IR LEDs to generate the transmitter's 'light-beam', and use matching IR photodiodes or phototransistors to detect the beam at the receiver end of the system. The graph of *Figure 3.20* conveys some useful information regarding the spectral response of the human eye and of general-purpose and IR photodiodes and phototransistors.

Thus, the human eye is sensitive to a range of electromagnetic light radiation; it has a peak spectral response to the colour green, which has a wavelength of about 550nm, and has relatively low sensitivity to the colour violet (400nm) at one end of the visible-light spectrum and to dark red (700nm) at the other; the human eye is blind to electromagnetic light radiation beyond this narrow spectrum. Optoelectronic semiconductor devices such as photodiodes and phototransistors have spectral responses that are determined by the chemistry of their semiconductor junction material; general-purpose 'light-sensitive' types have (as shown in *Figure 3.20*) typical spectral responses that straddle the human visibility spectrum, but IR types operate at a peak wavelength of about 900nm and generate an output spectrum that is well beyond the range of normal human visibility. IR light-beams are thus invisible to human eyes.

A simple IR direct-light-beam intrusion detector or alarm system can be made by connecting an IR light-beam transmitter and IR receiver in the basic way shown in *Figure 3.21*. Here, the transmitter feeds a coded pulse-type signal (often a simple squarewave) into an IR LED that has its output focused into a fairly narrow beam (via a moulded-in lens in the LED casing) that is aimed at a matching IR photodetector (phototransistor or photodiode) in the remotely placed receiver. The system action is such that the receiver output is 'off' while the light-beam reaches the receiver, but turns on and activates an external alarm or other mechanism if the beam is interrupted by a person or other object. This basic type of system can be designed to give a useful detection range of up to 20 metres when used with additional optical focusing lenses, or up to 5 metres without extra lenses.

Note that the simple *Figure 3.21* light-beam alarm system works on a strict line-of-sight principle between the Tx and Rx lenses, and the alarm may thus

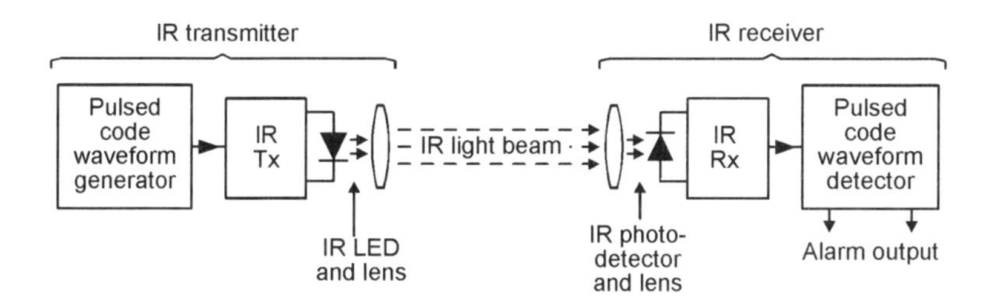

Figure 3.21 *Simple IR direct-light-beam alarm system*

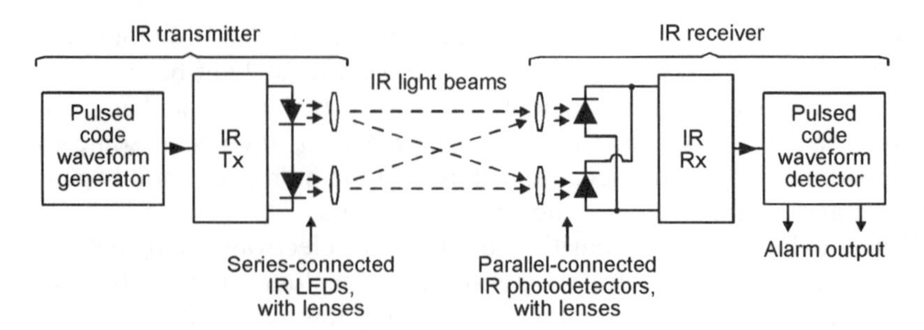

Figure 3.22 *IR dual-light-beam alarm system*

activate if any object with a diameter greater than the smaller of the two lenses enters the beam's line-of-sight. Thus, a weakness of this simple system is that it can easily be false-triggered by relatively small insects entering the beam or resting on one of the lenses. The improved dual-light-beam system shown in *Figure 3.22* does not suffer from this particular defect.

The *Figure 3.22* system is basically similar to that already described, but transmits the IR beam via two series-connected LEDs that are spaced about 75mm apart, and receives the beam via two parallel-connected photodetectors that are also spaced about 75mm apart. Thus, each photodetector can detect the beam from either LED, and the receiver's alarm will thus activate only if *both* beams are broken simultaneously, and this will normally only occur if a large (greater than 75mm) object is placed within the composite beam. This system is thus virtually immune to false triggering by insects, etc.

Note that, as well as giving excellent false-alarm immunity, the dual-light-beam system also gives (at any given LED drive-current value) double the effective detection range of the simple single-beam system, since it has twice as much effective IR transmitter output power and twice the receiver sensitivity.

IR system waveforms

IR light-beam systems are usually used in conditions in which high levels of ambient or background IR radiation (generated by natural or artificial heat sources) already exist. To enable the systems to differentiate against this background radiation and give good effective detection ranges, the transmitter beams are invariably pulse-coded, and the receivers are fitted with matching pulse-code detection circuitry. In practice, the transmitter beams usually use either a continuous-tone or a tone-burst type of pulse-coding, as shown in *Figure 3.23*.

IR LEDs and photodetectors are very fast-acting devices, and the effective range of an IR beam system is thus determined by the *peak* currents fed

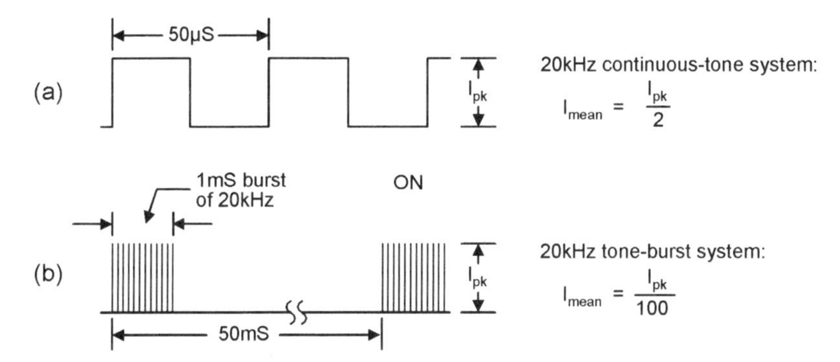

Figure 3.23 *Alternative types of IR light-beam pulse-code waveforms, with typical parameter values*

into the transmitting LED (or LEDs), rather than by the *mean* LED current. Thus, if the waveforms of *Figure 3.23* are used in IR transmitters giving peak LED currents of 100mA, both systems will give the same effective operating range, but the *Figure 3.23(a)* continuous-tone transmitter will consume a mean LED current of 50mA, while the tone-burst system of *Figure 3.23(b)* will consume a mean current of only 1mA (but will require a more complex circuit design).

The operating parameters of the tone-burst system require careful consideration, since this type of IR intrusion detecting system actually works on a 'sampling' principle and is usually intended to detect the presence of a human intruder. Note that humans moving at normal walking speed take about 200mS to pass any given point, so IR light-beam systems do not need to be switched on continuously to detect a human intruder, but only need to be turned on for brief 'sampling' periods at repetition periods that are far shorter than 200mS (at, say, 50mS); the actual sample period can be very short relative to the repetition period, but must be long relative to the tone frequency period. Thus, a good compromise is to use a 20kHz tone with a burst or 'sample' period of 1mS and a repetition period of 50mS, as shown in the waveform example of *Figure 3.23(b)*.

IR system design

The first step in designing any electronic system is that of devising the system's block diagrams. *Figure 3.24* shows a suitable block diagram of a continuous-tone IR light-beam intrusion alarm/detector system, and *Figure 3.25* shows the block diagram of a tone-burst version of the system. Note that a number of blocks (such as the IR output stage, the tone pre-amp, and the output driver) are common to both systems.

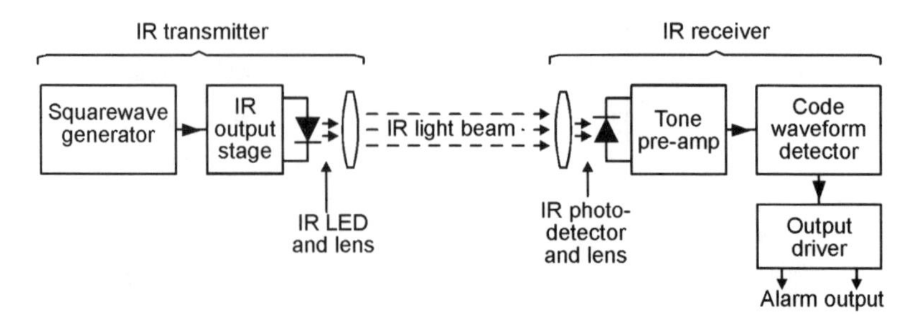

Figure 3.24 *Block diagram of a continuous-tone IR light-beam intruder alarm/detector system*

The continuous-tone system (*Figure 3.24*) is very simple, with the transmitter comprising nothing more than a squarewave generator driving an IR output stage, and the receiver comprising a matching tone pre-amplifier and code waveform detector, followed by an output driver stage that can activate a device such as a relay or alarm, etc.

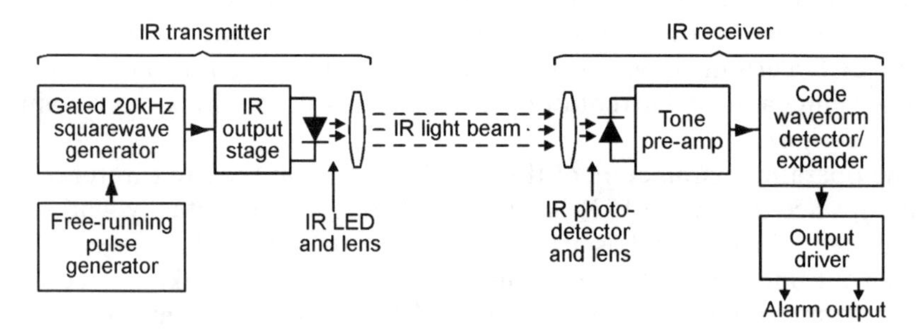

Figure 3.25 *Block diagram of a tone-burst IR light-beam intruder alarm/detector system*

The tone-burst system (*Figure 3.25*) is rather more complex, with the transmitter comprising a free-running pulse generator (generating 1mS pulses at 50mS intervals) that drives a 20kHz squarewave generator, which in turn drives the IR output stage that generates the final tone-burst IR light beam. In the receiver, the beam signals are picked up and passed through a matching pre-amplifier and then passed on to a code waveform detector/expander block, which ensures that the alarm does not activate during the 'blank' parts of the IR waveform. The output of the expander stage is fed to the output driver.

IR light-beam transmitter circuits

Figure 3.26 shows the practical circuit of a simple continuous-tone dual-light-beam IR transmitter. Here, a standard 555 'timer' IC is wired as an astable

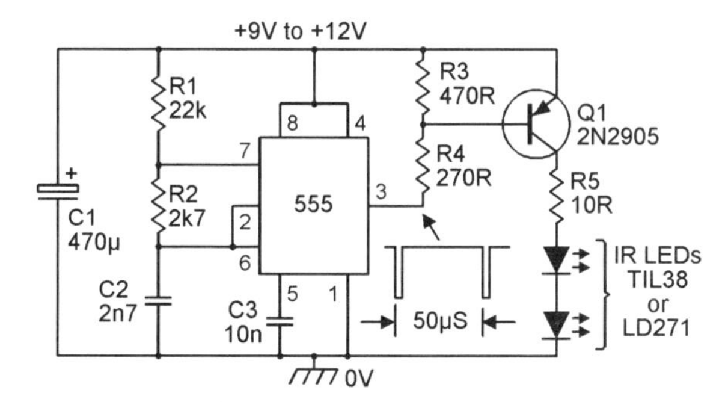

Figure 3.26 *Simple continuous-tone IR light-beam transmitter*

multivibrator that generates a non-symmetrical 20kHz squarewave output that drives the two series-connected IR LEDs at peak output currents of about 400mA via R5 and Q1 and the low source impedance of storage capacitor C1. The circuit's timing action is such that the ON period of the LEDs is controlled by C2 and R2, and the OFF period by C2 and (R1+R2), i.e. so that the LEDs are ON for only about one eighth of each cycle; the circuit thus consumes a *mean* current of only about 50mA.

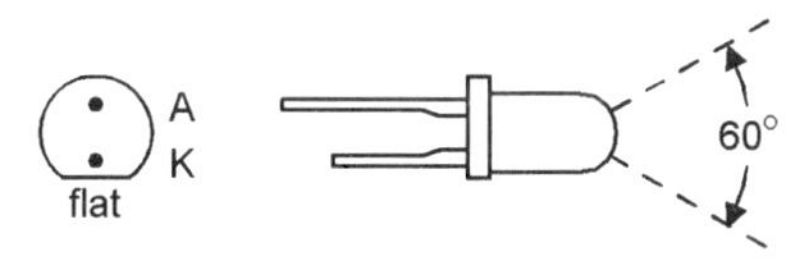

Figure 3.27 *Outline and connections used by the LD271 and TIL38 IR LEDs*

The *Figure 3.26* circuit can use either TIL38 or LD271 (or similar) 'high power' (100mW or greater) IR LEDs. These popular and widely-available LEDs can handle mean currents up to only 100mA or so, but can handle brief repetitive peak currents of up to at least 2.5A. *Figure 3.27* shows the outline and connections of these devices, which have a moulded-in lens that focuses the output into a radiating beam of about 60° width; at the edges of the beam the IR signal strength is half of that at the centre of the beam.

Minor weaknesses of the IR output stage (Q1 and R3 to R5) of the *Figure 3.26* circuit are that it has a very low input impedance (about 300R), that it gives an inverting action (the LEDs are ON when the input is low), and that the LED output current varies with the circuit's supply voltage. *Figure 3.28* shows an alternative universal IR transmitter output stage that suffers from none of these defects.

In *Figure 3.28*, the base drive current of output transistor Q2 is derived from Q1 collector, and the Q1 circuit has an input impedance of about 5k0

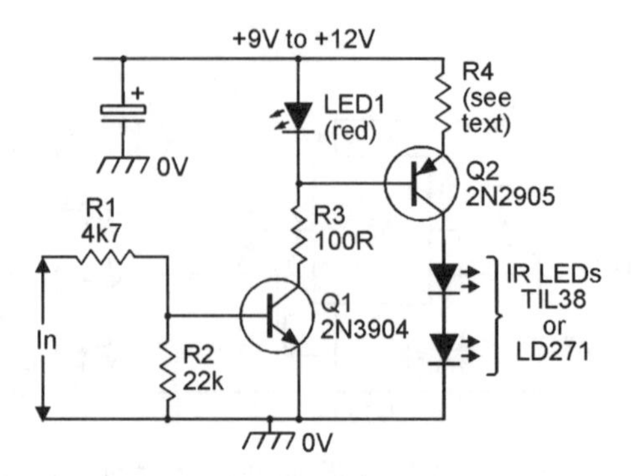

Figure 3.28 *'Universal' IR transmitter output stage*

(determined mainly by the R1 value). Thus, when the circuit's input is low Q1 is off, so Q2 and the two IR LEDs are also off, but when the input is high Q1 is driven to saturation via R3, thus driving LED1 (a standard red LED) and Q2 and the two IR LEDs on. Note that under this latter condition about 1.8V is developed across LED1, and that about 0.6V less than this (= 1.2V) is thus developed across R4, causing Q2 to act as a constant-current generator that feeds a peak collector current of 1.2V/R4 amps into the two IR LEDs. Thus, this circuit's peak output current can be set by giving R4 an ohms value of 1.2V/I, where I is the desired peak output current in amps.

Figure 3.29 shows a 20kHz squarewave generator (made from a 555 timer IC) that can be used in conjunction with the *Figure 3.28* output circuit to make a continuous-tone IR beam transmitter. In this case the *Figure 3.28* circuit's R4 value should be at least 6R8, to limit the peak IR LED currents to less that 200mA.

Alternatively, *Figure 3.30* shows the circuit of a tone-burst generator that gives 1mS bursts of 20kHz at 50mS intervals and which can be used in conjunction with the *Figure 3.28* output stage to make an IR tone-bursts

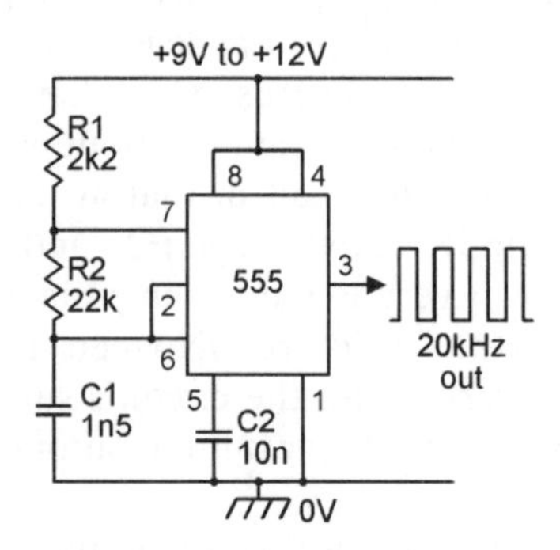

Figure 3.29 *20kHz squarewave 'tone' generator*

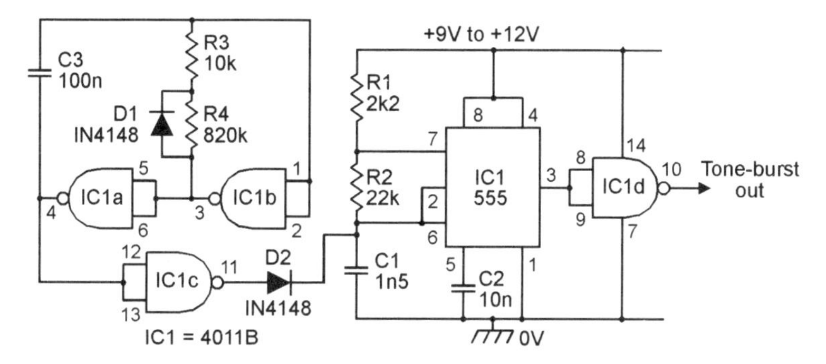

Figure 3.30 *Tone-burst (1mS burst of 20kHz at 50mS intervals) waveform generator*

transmitter. Here, the IC1a and IC1b sections of a 4011B CMOS quad 2-input NAND gate IC are wired as a free-running asymmetrical astable multi-vibrator that produces 1mS and 49mS periods; this waveform is inverted and buffered by IC1c and used to gate a 20kHz 555-type squarewave generator via D2, and this squarewave is then buffered and inverted by the final 4011B stage (IC1c), ready for feeding to the input of the *Figure 3.28* output stage.

Note when using the *Figure 3.30* circuit that R4 in the *Figure 3.28* output stage can be given a value as low as 2R2, to give peak output currents of up to 550mA, but that under this condition the transmitter will consume a mean current of little more than 6mA.

IR receiver pre-amp design

The basic IR input signal to an IR light-beam receiver can be picked up and converted into a proportional current by either an IR photodiode or an IR phototransistor. If a photodiode is used, it can be connected in series with a load resistor (with a typical value in the range 10k to 100k) and used in either

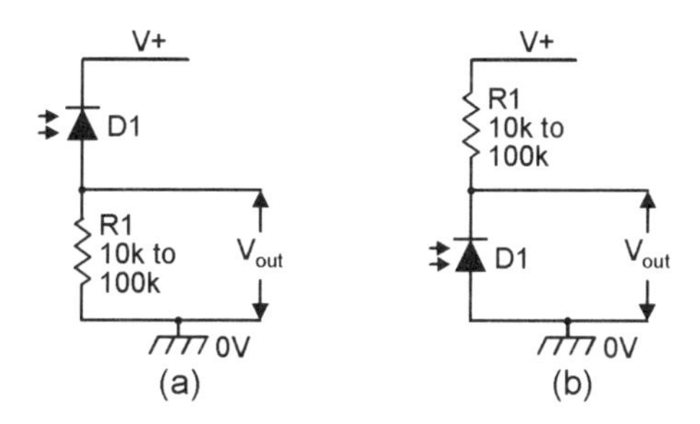

Figure 3.31 *Alternative ways of using a photodiode as a light-to-voltage converter*

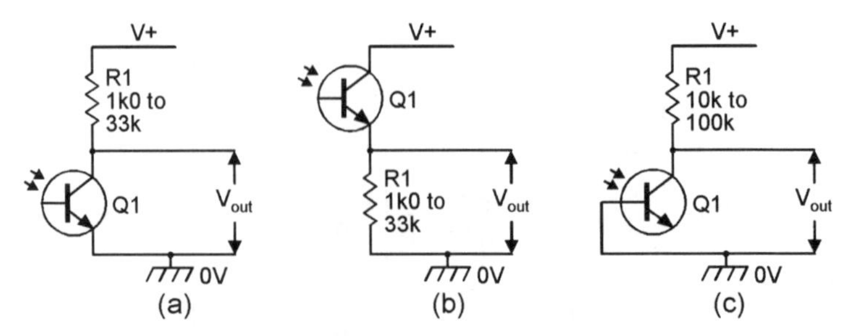

Figure 3.32 *(a) and (b); alternative phototransistor configurations. (c); a 3-lead phototransistor used as a photodiode*

of the reverse-biased configurations shown in *Figure 3.31*. The diode's basic action is such that its reverse-biased leakage current is proportional to the IR light intensity on its junction, being very low under dark conditions and relatively high when brightly illuminated; this current is converted into a proportional output voltage by R1.

An IR phototransistor can be used by connecting it in either of the basic ways shown in *Figure 3.32(a)* or *(b)*, in which load resistor R1 has a typical value in the range 1k0 to 33k. Most phototransistors have only two externally-accessible leads (collector and emitter), but a few are 3-lead types with an accessible base lead; a 3-lead device can be used as a phototransistor by connecting it in either of the two basic ways already shown, or can be used as a photodiode by connecting it in the way shown in *Figure 3.32(c)*.

Note that a phototransistor's sensitivity is typically one hundred times greater than that of a photodiode, but its maximum operating frequency (typically a few hundred kHz) is proportionally lower than that of a photodiode (typically tens of MHz). Also note in *Figures 3.31* and *3.32* that the photosensor exhibits a high sensitivity but a low cut-off frequency if R1 has a high value, and a low sensitivity but high cut-off frequency if R1 has a low value.

Figure 3.33 shows the practical circuit of an IR light-beam receiver that is designed for use with 20kHz continuous-tone or tone-burst single-beam or dual-beam systems, and using IR photodiodes as signal converters. Here, the two IR diodes are connected in parallel and wired in series with R1, so that the converted IR signal is developed across R1 (note that one of these diodes can be removed if the unit is used with a single-beam IR system). The converted R1 signal is amplified by cascaded op-amps IC1 and IC2, which can provide a maximum signal gain of about ×17 680 (= ×83 via IC1 and ×213 via IC2), but have the gain made variable via RV1. These two amplifier stages have their frequency responses centred on 20kHz, with third-order low-frequency roll-off provided via C4–C5–C6 and with third-order high-frequency roll-off provided by C3 and the internal capacitors of the two op-amps.

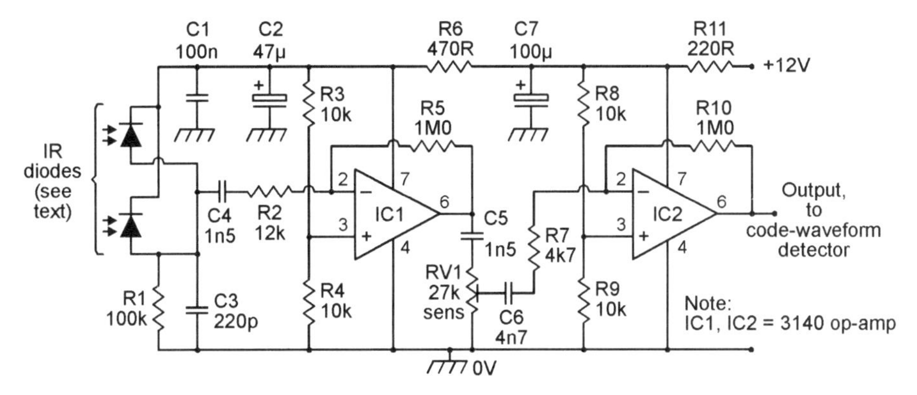

Figure 3.33 *IR receiver pre-amplifier circuit*

The *Figure 3.33* receiver pre-amp circuit can be used with a variety of IR detector diode types, which ideally should be housed in black (rather than clear) infra-red transmissive mouldings, which greatly reduce unwanted pick-up from visible light sources. *Figure 3.34* shows the case outline and IR-sensitive face positions of three popular IR photodiodes of this type.

The output of the *Figure 3.33* pre-amplifier can be taken from IC2 and fed directly to a suitable code waveform detector circuit, such as that shown in *Figure 3.35*. Note, however, that if the IR Tx–Rx light-beam system is to be used over ranges less than 2 metres or so, the pre-amp output can be taken directly from IC1 and all the RV1 and IC2 circuitry can be omitted from the pre-amp design.

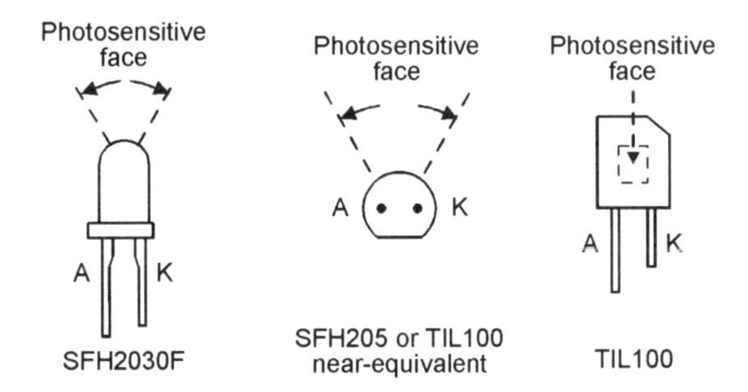

Figure 3.34 *Case outline and IR-sensitive face positions of three popular types of IR photodiode*

A code waveform detector

In the *Figure 3.35* code waveform detector circuit the 20kHz tone waveforms (from the pre-amp output) are converted into dc via the D1–D2–C2–R5–C3

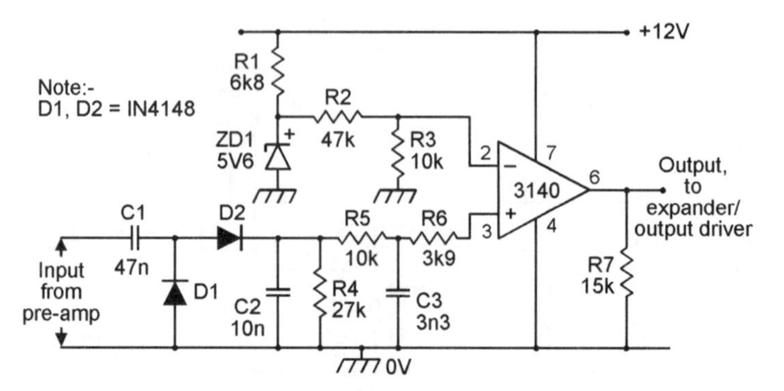

Figure 3.35 *Code waveform detector circuit*

network and fed (via R6) to the non-inverting input of the 3140 op-amp voltage comparator, which has its inverting input connected to a thermally stable 1V0 dc reference point. The overall circuit action is such that the op-amp output is high (at almost full positive supply rail voltage) when a 20kHz tone input signal is present, and is low (at near-zero volts) when a tone input signal is absent; if the input signal is derived from a tone-burst system, the output follows the pulse-modulation envelope of the original transmitter signal. The detector output can be made to activate a relay in the absence of a beam signal by using the expander/output driver circuit of *Figure 3.36*.

An expander/output driver

The operating theory of the *Figure 3.36* circuit is fairly simple. When the input signal from the detector circuit switches high C1 charges rapidly via D1, but when the input switches low C1 discharges slowly via R1 and RV1;

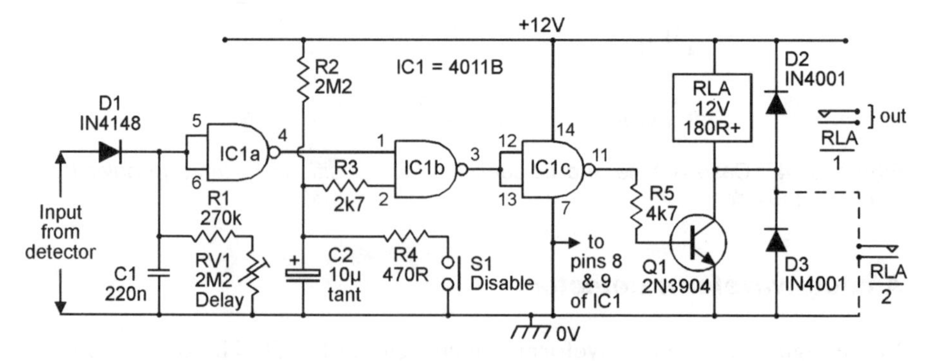

Figure 3.36 *Expander/output driver circuit*

C1 thus provides a dc output voltage that is a 'time-expanded' version (with expansion presettable via RV1) of the dc input voltage. This dc output voltage is buffered and inverted via IC1a and used to activate relay RLA via Q1 and an AND gate made from IC1b and IC1c.

Normally, the other (pin-2) input of this AND gate is biased high via R2, and the circuit action is such that (when used in a complete IR light-beam system) the relay is off when the beam is present, but is driven on when the beam is absent for more than 100mS or so. This action does not occur, however, when pin-2 of the AND gate is pulled low; under this condition the relay is effectively disabled.

The R2–C2 network's purpose is to disable the relay network via the AND gate (in the way just described) for several seconds after power is initially connected to the circuit or after DISABLE switch S1 is briefly operated, thus enabling the owner or other authorized persons to pass through the beam without activating the relay. Note that the relay can be made self-latching, if required, by wiring normally-open relay contacts RLA/2 between Q1 emitter and collector, as shown dotted in *Figure 3.36*.

IR light-beam system ranges

The circuits of *Figures 3.33*, *3.35*, and *3.36* can be directly interconnected to make a complete IR light-beam receiver that can respond to either tone-burst or continuous-tone signals; the receiver must be powered from a well-regulated 12V DC supply unit, such as that shown in *Figure 3.37*. The practical maximum operating range of a complete IR light-beam security system of this type is greatly affected by the types of lenses used in the system. If additional lenses are *not* used, but the Tx and Rx are carefully aimed at each other, and the Rx diodes are screened from the effects of visible light by mounting them deep inside aimed tubing, the maximum range should be at least 5 metres, and may be as high as 7 metres. This range can

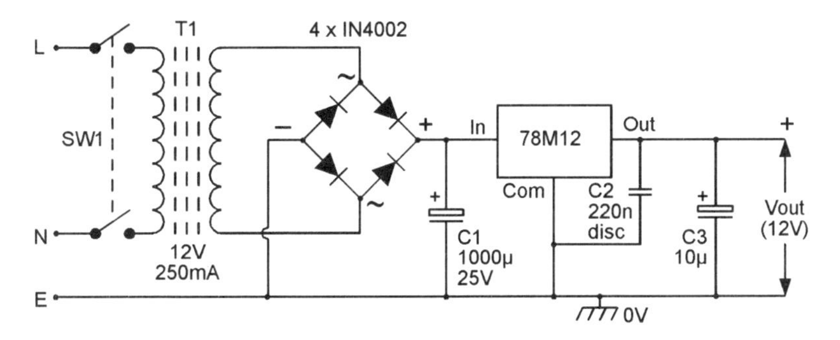

Figure 3.37 *Line-powered regulated 12V, 250mA supply*

be vastly increased with the help of additional focusing lenses and/or reflectors.

At the Tx end of the system, most of the optical output power of each IR LED is typically radiated over an arc of about 60°, and thus has a fairly low radiation density value; the Tx signal's radiation density value can easily by increased by a factor of four (thus doubling the system's effective range) by mounting each IR LED at the focal point of a simple torch-type optical reflector that is aimed at the receiver unit.

At the Rx end of the light-beam system, each IR photodiode has an integral lens that focuses the received IR light on to the diode's photosensitive area. On 5mm detectors such as the SFH2030F, this lens has a collection area of about 19.6mm^2; in this example, the detector's effective sensitivity can be increased by a factor of four (thus doubling the system's range) with the help of an external 10mm^2 focusing lens, or by a factor 36 (thus increasing the range by a factor of six) with the help of a 30mm^2 focusing lens. Thus, if reasonable care is taken in the opto-mechanical design of the IR system, its range can easily be increased to 20 metres, and possibly to 60 metres or more.

PIR movement-detecting systems

IR light-beam alarms are 'active' IR units that react to an artificially generated source of IR radiation. Passive IR (PIR) alarms, on the other hand, react to naturally generated IR radiation such as the heat-generated IR energy radiated by the human body, and are widely used in modern security systems. Most PIR security systems are designed to activate an alarm or floodlight, or open a door or activate some other mechanism, when a human or other large warm-blooded animal moves about within the sensing range of a PIR detector unit, and use a pyroelectric IR detector of the type shown in *Figure 3.38* as their basic IR-sensing element.

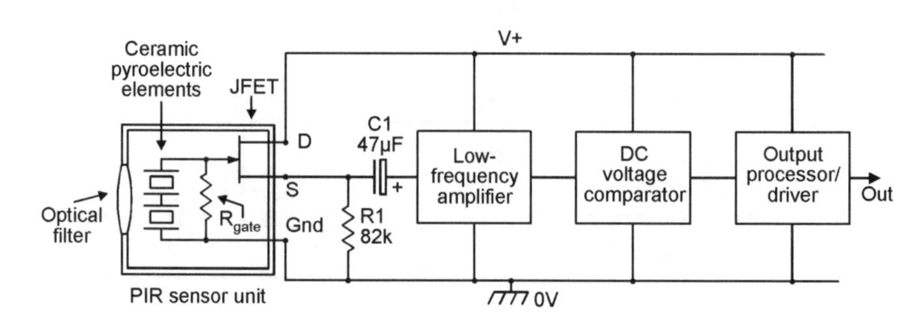

Figure 3.38 *Basic PIR detector usage circuit*

The basic *Figure 3.38* pyroelectric IR detector relies on the fact that some special ceramics generate electrical charges when subjected to thermal variations or uneven heating. Modern pyroelectric IR detectors such as the popular PIS201S and E600STO types incorporate two small opposite-polarity series-connected ceramic elements of this type, with their combined output buffered via a JFET source-follower, and have the IR input signals focused onto the ceramic elements by a simple filtering lens, as shown in the basic PIR detector usage circuit of *Figure 3.38*. It is important to note at this point that the detector's final output voltage is proportional to the *difference* between the output voltages of the two ceramic elements.

The basic action of the *Figure 3.38* PIR detector is such that, when a human body is within the visual field of the pyroelectric elements, part of that body's radiated IR energy falls on the surfaces of the elements and is converted into small but detectable variation in surface temperature and corresponding variation in the output voltage of each element. If the human body (or other source of IR radiation) is stationary in front of the detector's lens under this condition, the two elements generate identical output voltages and the unit's final 'difference' output is thus zero, but if the body is moving while in front of the lens the two elements generate different output voltages and the unit produces a varying output voltage.

Thus, when the PIR unit is wired as shown in the *Figure 3.38* basic usage circuit, this movement-inspired voltage variation is made externally available via the buffering JFET and dc-blocking capacitor C1 and can, when suitably amplified and filtered, be used to activate an alarm or other mechanism when a human body movement is detected. In practice, pyroelectric IR detectors of the simple type just described have, because of the small size (usually about 20mm^2) and simple design of the detector's IR-gathering lens, maximum useful detection ranges of roughly one metre. In modern commercial PIR movement detecting security units, however, this range is greatly extended (usually to well over ten metres) with the aid of a large (about 2000mm^2) multi-faceted external IR-gathering/focusing plastic lens, which splits the visual field into a number of parallel strips and focuses them onto the two sensing areas of the PIR unit.

Figure 3.39 shows the typical PIR sensing pattern of a commercial 'intrusion detector' unit designed to protect a normal-sized room in domestic-type applications. In this example the unit is mounted on a wall at a height of seven feet and is aimed downwards at a shallow angle, and the multi-faceted plastic lens splits the visual field into a large number of vertical and horizontal segments. Any person moving through a single segment will activate a single trigger signal within the PIR sensor; a person moving through the entire visual field will thus produce numerous triggering signals, but a stationary IR source will produce no signals. Most intrusion detectors of this type incorporate 'event counting' circuitry that will only generate an alarm-activating output if three or more trigger signals are detected within a few seconds, thus

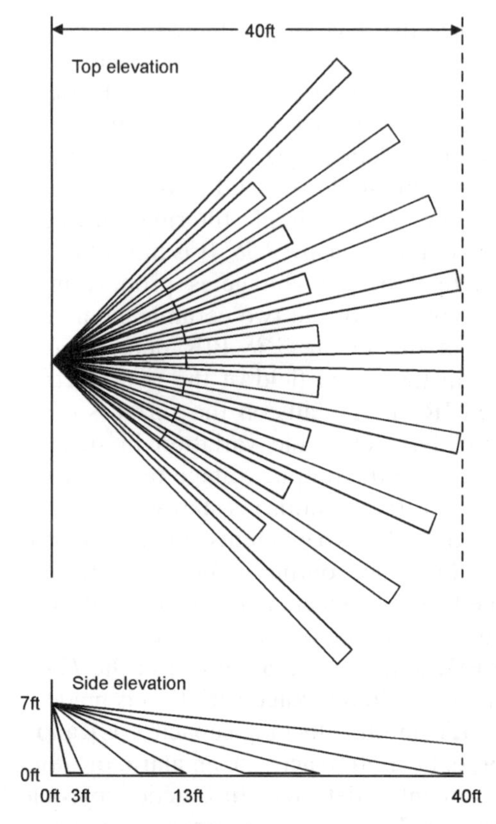

Figure 3.39 *Typical PIR sensing pattern of a commercial 'intrusion detector' unit designed for normal domestic-type applications*

minimizing the chances of a false alarm due to sudden changes in temperature caused by the auto-activation of time-switched security lights, etc.

The lens-generated PIR sensor pattern shown in *Figure 3.39* is the type that is often used in burglar-alarm systems to protect a single room in a medium-sized house. Alternative (and usually interchangeable) plastic lens types, offering different ranges and coverage patterns for various special types of application, are available at low cost from many commercial PIR-unit suppliers. Among the most important of these are the 'pet' type, in which the field's vertical span is restricted to 2.5 to 6.6 feet above ground level to avoid activation by domestic pets while giving good sensitivity to normal humans, and the 'corridor' type, in which the field's horizontal span is restricted to about 20 degrees to give long-distance coverage (typically about 30 metres) of narrow corridors and passageways.

Note that, because high-quality commercial PIR security units of this basic type are widely available at comparatively low cost, it is not practicable (on aesthetic and cost-effective grounds) to try to build similar units on a DIY basis.

4

Anti-burglary security circuits

The best known and most widely used type of electronic security unit is the so-called 'burglar alarm', which is usually designed to detect burglars attempting to break into protected premises while at the same time letting authorized individuals roam freely within those premises and to easily enter or leave their perimeter zones. It is important to understand that normal burglar alarms are last-resort devices that cannot be used on a 'stand alone' basis to give good protection against burglary; to give good protection, they must always be used in conjunction with various physical defences/deterrents and with mechanical security devices such as locks, latches, and security windows, and their sensors and sirens must be located to give the best possible security to the individual premises that are being protected. Note that the main aim of an anti-burglary security *system* is that of deterring potential burglars from even *trying* to break into the defended premises; the burglar alarm is simply a back-up to that system and activates when a burglary is actually in progress, and is thus of value only if the main deterrent security system fails in its task.

To get the best value from any burglar alarm the user must first learn the basic principles of anti-burglary protection, and must then use that knowledge to convert the basic burglar alarm into a device that gives a performance that is tailored to suit some specific security application. This chapter gives a concise outline of these various principles, and then goes on to show a variety of practical domestic-type burglar alarm and accessory circuits.

Anti-burglary basics

The burglar

A burglar is a person who forcibly enters houses or other premises with the intention of theft. To be burgled is a vile experience. At best, the burglar may be a professional who will enter your home and steal many of your personal possessions, some of which will have a sentimental value far in excess of their insured monetary cost; the next day that same burglar will probably sell those precious goods for a trivial amount of money, and then go and rob someone else. At worst, your burglar may be a demented amateur who enters your home with a heart full of hate, intent on stealing your cash and destroying or desecrating everything else; he will slash your furniture and clothes, urinate on your bedding, smear excrement and paint on your walls, and try to smash or burn everything else; this type of burglary is so repulsive that many victims never recover from the psychological damage caused by the experience.

Your vulnerability to burglary is greatly influenced by the location and nature of the premises in which you live or work, and by the security precautions that you take to protect those premises. In all advanced Western countries, annual burglary totals are proportional to national population figures, and are typified by those of the UK, which has a total population of about 56 000 000. In the UK, the annual total of burglaries in the ten years up to 1997 averaged about 900 000, of which approximately 400 000 were domestic burglaries. Of these domestic incidents, almost three quarters involved actual physical forced entry, and more than a quarter were walk-in burglaries in which the intruder entered the premises via an unlocked door or window or (in a small number of cases) by using a carelessly hidden spare front door key.

For security purposes, all private premises can be regarded as defensible fortresses (houses or other buildings) that are surrounded by defensible outer border zones (private land, with outer fences and hedges, etc.). No burglar can reach your private fortress without first passing through at least one of its outer border zones, which thus form your first and most important anti-burglary defence areas; if you use these areas sensibly, you will deter most burglars from even *trying* to break into your house. *Figure 4.1* and the next two major sections of text help illustrate some of the basic defence principles of these areas, specifically applied to medium-sized mid-terrace and semi-detached houses of the types found in many suburban areas in the UK.

Front outer border defence zones

Most burglars make an unobtrusive visual study of a building by walking or driving casually past it to determine whether or not it seems an easy or a not-worth-the-risk target for burglary. Their main aim is to get into and out of the premises unobtrusively, and they thus like houses that have their front gardens enclosed by high shrubs or bushes, or are shrouded in darkness at night, and seem to be unoccupied and to have no obvious anti-burglary protection. Many potential burglars can be scared off by taking simple

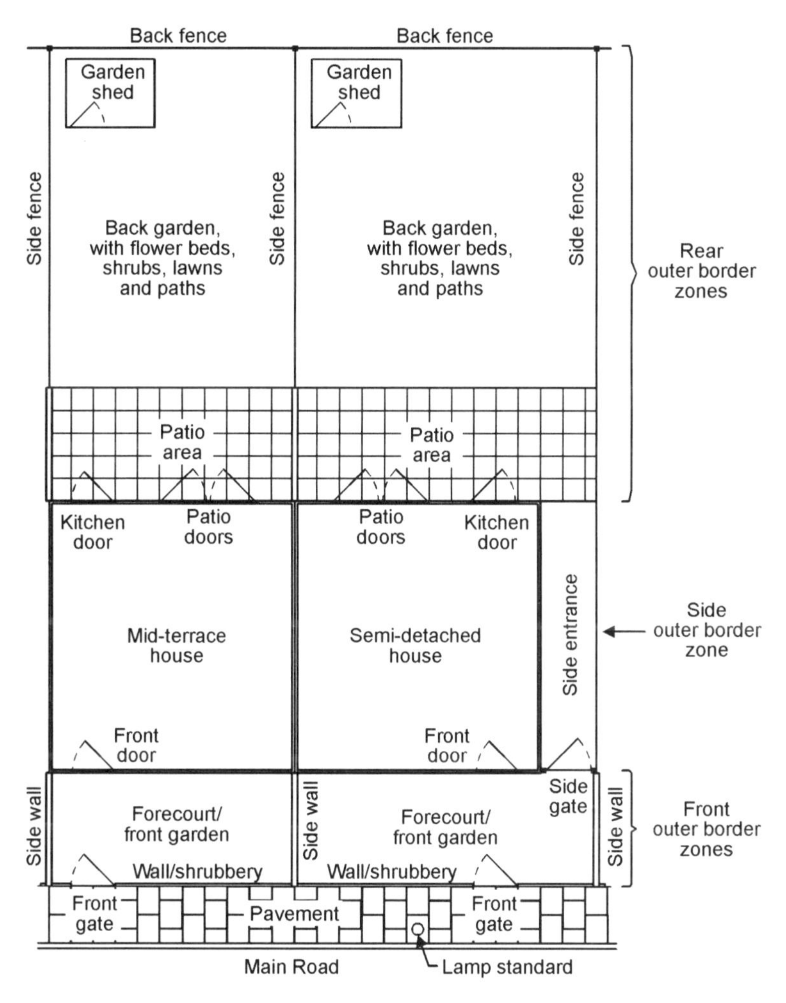

Figure 4.1 *Diagram showing the outer border defence zones of typical UK-style mid-terrace and semi-detached houses (see text)*

precautions such as fitting the building with time switches that automatically operate house lights at various times, or by leaving a radio switched on when the house is empty, or by fitting a real or dummy alarm bell housing to the front of the house.

One priceless anti-burglary device is the nosey but trustworthy neighbour who, if asked, will keep an eye on your house front when you are away on holiday, and will immediately summon help if anything suspicious is seen. Treat such people kindly, and try to gain their friendship.

If burglars decide to attack a house from the front, they usually enter the garden via the front gate and then try to enter the house by the front door

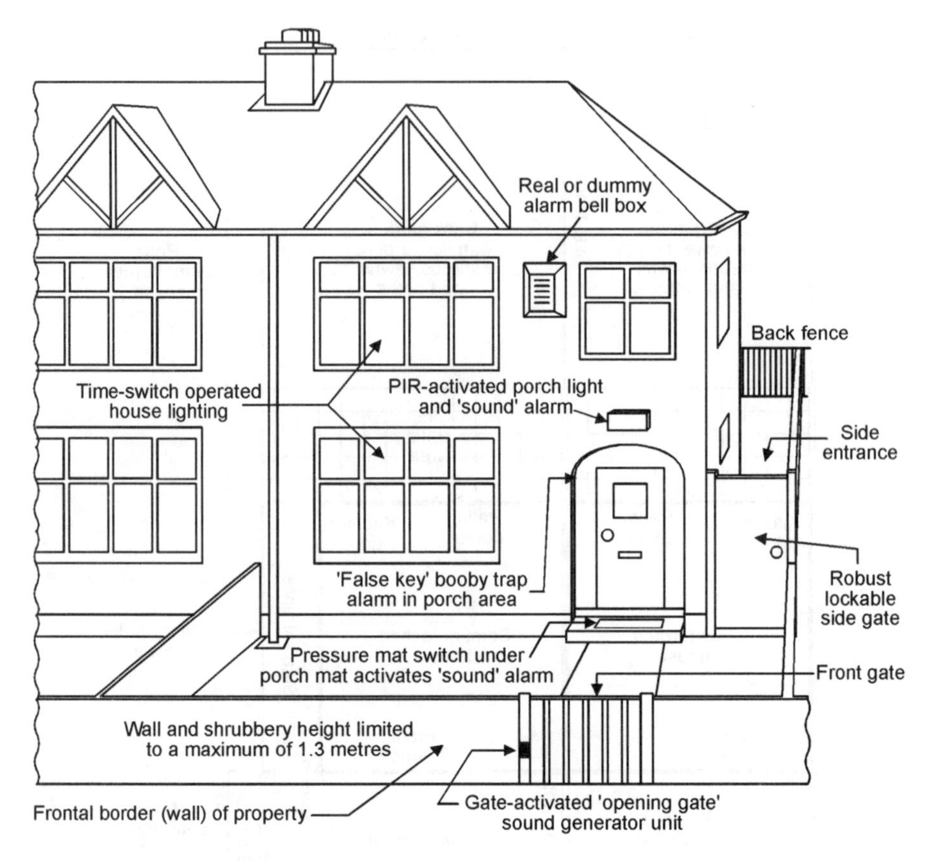

Figure 4.2 *Diagram showing various burglar deterrents used in the front outer border zone of a semi-detached house*

or – if the house has a side entrance – by forcing the side gate and then entering the house from the rear. Burglars know that some people hide a spare front door key under a flower pot or porch mat or on a porch ledge when they leave the house, and burglars often make a quick search for such a key when they enter the porch. You can take advantage of this fact by making a simple 'false key' booby trap that activates a self-latching alarm if an object such as a flower pot is briefly moved, or if someone grabs a key that is tied to a microswitch by a short length of string, etc. Thus, basic rules for protecting the house against frontal attack are as follows, and are illustrated in *Figures 4.2* and *4.3*:

(1) Limit all front garden shrubbery to a maximum height of 1.3 metres.
(2) If your house front is poorly lit at night, fit it with an automatic PIR-activated porch lighting system.

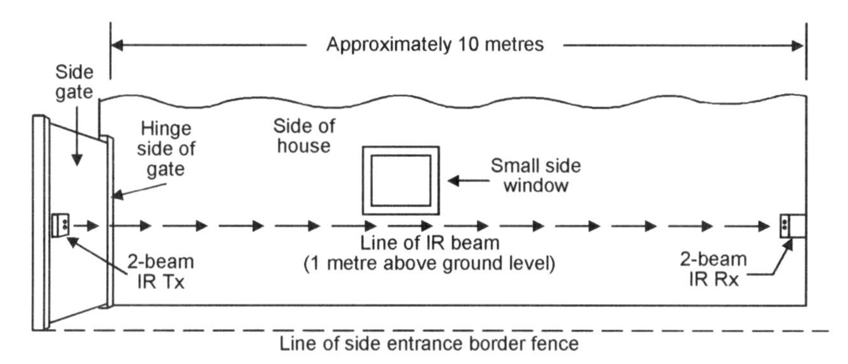

Figure 4.3 *Diagram showing basic arrangement of an IR light-beam side entrance defence system*

(3) If your front gate is a self-closing type, fit it with a device that generates a distinctive sound as the gate is opened.
(4) If your front porch is not fitted with automatic PIR-activated lighting but does have a porch mat, place a pressure mat switch under the porch mat and use it to activate some clearly audible or visual warning device when anyone treads on the porch mat.
(5) Deter burglars by fitting a real or dummy alarm bell box to the front of the house, by fitting the house with time-switch operated lighting, and by leaving a radio turned on when the house is unoccupied.
(6) Fit a simple 'false key' booby trap in the porch area.
(7) If you have a side entrance, make sure that it is fitted with a robust and lockable side gate, at the house-front end of the entrance. Fit the side entrance with some type of intrusion-detecting alarm device that will activate if any unauthorized person enters the side entrance (even if they do so by climbing the gate), but will not be activated by cats, etc. House side entrances have a typical length of about 10 metres and are best defended by a PIR alarm fitted with a 'corridor' type of lens, or by an IR dual 'light-beam' alarm of the type described in Chapter 3 of this book; *Figure 4.3* shows the basic arrangement of a suitable 2-beam IR light-beam alarm system.

In *Figure 4.3* the IR dual-beam transmitter (Tx) is fixed to the side gate, near its 'fence' edge, and is aimed diagonally along the length of the side entrance, at a height of about 1 metre, towards the IR receiver (Rx) unit, which is fixed to the house wall at the far end of the entrance. Thus, the beam's Tx to Rx contact will be broken if the gate is opened, or if any adult person moves along the side entrance when the gate is closed. Note that in this 'slow movement' type of application the system will work adequately if the Tx unit sends a 1mS high-frequency test pulse once every 200mS or so;

thus, if this test pulse has a peak amplitude of (say) 100mA, the Tx unit may consume a mean current of only 250μA. The Tx and Rx units can thus easily be continuously operated and powered by rechargeable battery units that are trickle charged (from a single mains-powered unit) via a pair of low-voltage low-current leads.

Rear outer border defence zones

Three-quarters of all house break-ins occur at the back of the house (where the burglar is least likely to be seen), usually via a window. To reach the house, the burglar must cross the garden and patio areas, and usually reaches these by climbing a back or side fence or via a side entrance. Houses with rear access via a shared passage or driveway are particularly vulnerable to this type of break-in. Once a burglar has broken into one house, he can easily gain access to adjacent houses by climbing over their side fences, sometimes breaking into several houses or their sheds or garages in a short space of time. Thus, basic rules for protecting the house against attack from the rear are as follows, and in some cases are illustrated in *Figure 4.4*:

(1) Burglars treat garden sheds and garages as valuable sources of tools and equipment for the break-in and for later sale, and often break into them before attacking the actual house. Thus, always fit your shed and garage with a simple battery-powered burglar alarm that activates a loud siren and a light strobe if not disabled (via a secret switch) within about 20 seconds of opening the shed/garage door.
(2) If you keep a ladder in the rear defence area, wire it to a simple 'loop' alarm (see Chapter 2), so that the alarm sounds if the ladder is forcibly moved.
(3) If your house has rear access via a passage or driveway, protect the top of your back fence/wall with barbed wire or in some other way.
(4) It is not practical to protect side fences against a burglar who wants to climb over them. Usually, however, the burglar is climbing them to reach the back of your house via the patio area, which can easily be protected via a PIR-activated flood-light/alarm unit or an IR light-beam alarm unit that is aimed along the back of the house a metre or so above ground level, as shown in *Figure 4.4*. If an IR light-beam is used, it must be positioned so that its beam can not be broken by carelessly placed patio furniture or by growing shrubbery.
(5) Sometimes the burglar may climb your garden fences to reach an adjacent property. You can detect this type of intrusion with an IR light-beam alarm unit that is aimed along the length of the garden, between the house and the garden shed (see *Figure 4.4*).

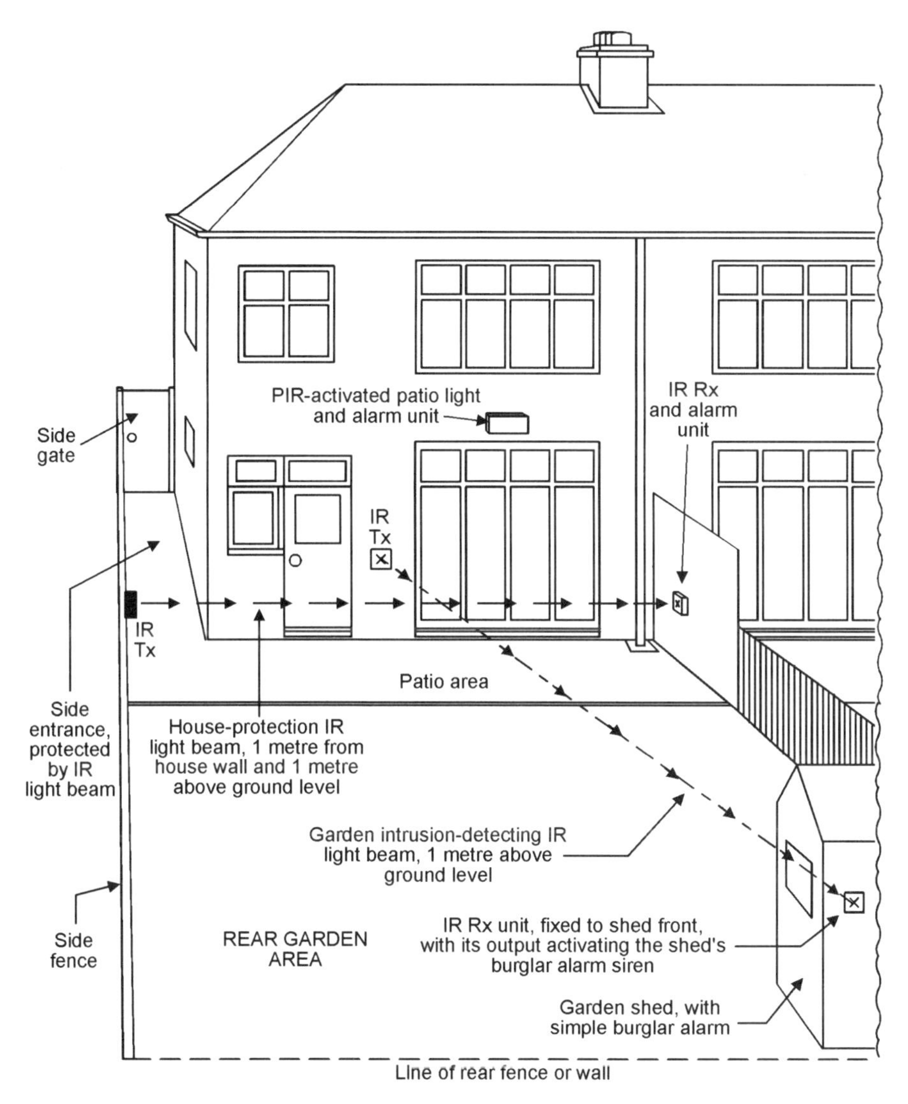

Figure 4.4 *Diagram showing various burglar deterrents used in the rear border zone of a semi-detached house*

(6) If your house has a shared side entrance that leads to a pair of garages, mount a cheap dummy TV camera (with a built-in flashing LED that is powered from a remote battery) in a hard-to-reach position on the apex of the garages, aiming it along the entrance so that it is clearly visible from the street, as a burglar deterrent.

Front-door robbers

The easiest way to get into someone's house is to simply knock on its front door and, when it opens, either barge or trick your way into the premises. Small-time crooks (including small children) often use the latter technique to carry out petty robberies, typically arriving as a pair and getting into the house on a flimsy pretext such as using your phone to make an emergency call, or asking for a glass of water, etc. One of the pair then keeps you busy with idle chit-chat while the other person searches the house for loose cash and trinkets. The basic rules for protecting yourself against this type of robbery are as follows:

(1) Fit your front door with a security chain, and never unhook it unless you are sure it is safe to do so.
(2) Buy a fixed or mobile self-latching panic-button alarm, and keep your finger on its button whenever you open the front door to a stranger.
(3) Never allow *anyone* (including children) into your house unless you are absolutely sure it is safe to do so.
(4) Never, under any circumstances, allow two or more strangers (particularly innocent looking children) into your house at the same time.
(5) Never, ever, under any circumstances, leave a total stranger alone in any part of your house; if they have the impertinence to ask to be left alone (to make a 'private' telephone call), immediately order them to leave the house, sound your panic alarm, and – as soon as they leave – call the local police.

The house

If the would-be burglar has successfully passed through your building's outer border defence zones, he will now start to do physical and costly damage to the actual building by trying to break into it. His chances of breaking in are reduced (but the cost of the inflicted damage is increased) if the house is fitted with strong outer doors with robust locks/latches, and with double-glazed self-locking windows. You can greatly reduce the crook's chances of committing a successful burglary by fitting the house (or other building) with a properly designed burglar alarm system.

Any building can, for crime prevention purposes, be regarded as a box that forms an enclosing perimeter around a number of interconnected compartments. This 'box' is the shell of the building, and contains walls, floors, ceilings, doors and windows. To commit any crime within the building, the intruder must first break through this shell, which thus forms the owner's first line of house defence. In most houses, the most vulnerable parts of the shell are its doors and windows, but ceilings are also vulnerable (often via a

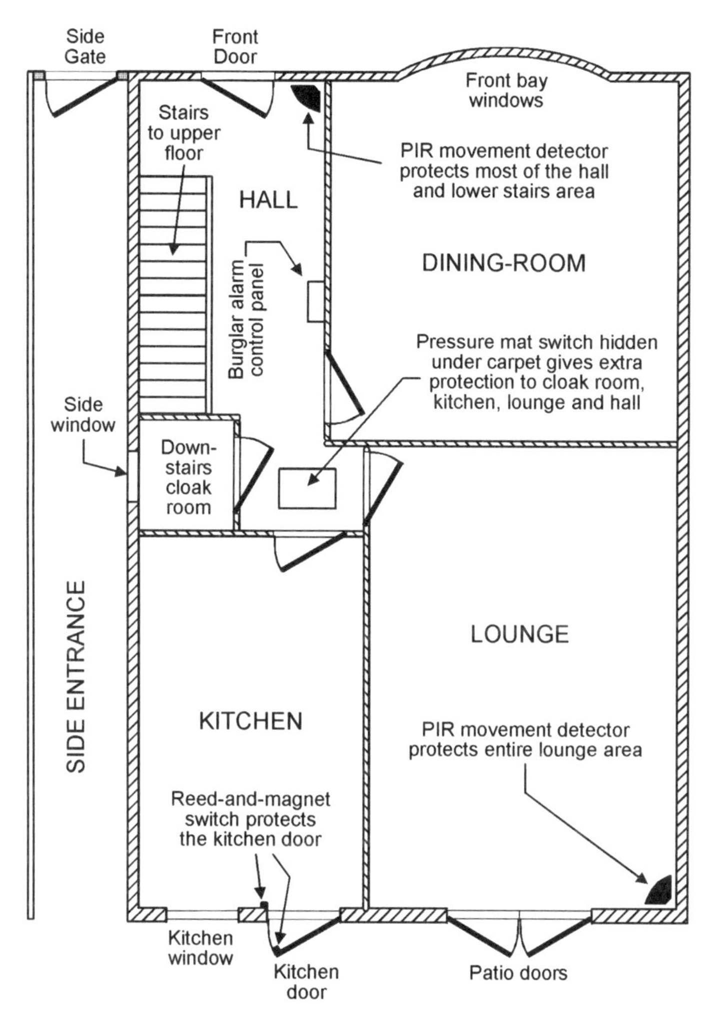

Figure 4.5 *Ground-floor plan of a semi-detached house, showing suitable positions for anti-burglary defences*

trap door and loft space) in some top-floor flats and apartments and in many commercial buildings.

Once an intruder has entered the building, he can move from one room or compartment to the next only along paths that are predetermined by the layout of internal doors and passages. In moving from one compartment to the next he must inevitably pass over or through certain spots or areas in the building, as is made clear in *Figure 4.5*, which shows the ground-floor plan of a medium-sized house, together with suitable positions for anti-burglary devices/sensors such as PIR movement detectors, pressure mat switches, and reed-and-magnet 'contact' switches.

Thus, if an intruder breaks into this house via the kitchen (which has its outer door protected by a reed-and-magnet contact switch), or via the downstairs cloak room (via the side window), he can only reach the rest of the house by entering the hall, which is well protected by a hidden pressure mat switch and by a PIR movement detector, which also protects the front door and most of the stairway. The entire lounge is well protected by another PIR unit, which is aimed away from direct sunlight (a common cause of false alarms in PIR units) but will respond instantly if anyone enters the room via its main door or its patio doors. The dining-room is not individually protected, since it can only be accessed by its doorway (which is protected by the hall's PIR unit), or via its bay window, which (if the windows are clearly visible from the street) is a very unlikely attack point. Thus, the entire ground floor of this house is adequately protected by just two PIR units and two detector switches. The upper floor can be protected with similar simplicity.

Note in *Figure 4.5* that the burglar alarm's main control panel is situated in the hall, where it can be conveniently operated (via a security key) on entering or leaving the house, or prior to going up the stairs (to go to bed), or immediately after coming down the stairs (after getting out of bed). The basic operating details of the burglar alarm system and its control panel are described in the next major section of this chapter.

Burglar alarm basics

The burglar alarm system

Most modern domestic burglar alarm systems consist of a number of switched-output intrusion sensors (contact switches and PIR units, etc.) and a 'panic' button, which have their outputs coupled to the inputs of a master control unit that processes the received signals and, when appropriate, activates a built-in medium-power audible alarm and can, if required, also activate a high-power external siren or bell. The burglar alarm system's block diagram thus takes the basic form shown in *Figure 4.6.*

Note in *Figure 4.6* that the alarm sensors are arranged in blocks or groups, each of which is allocated a specific 'zone' input connection point on the main control unit. The basic idea here is to divide your property into a number of distinct defence zones, each of which can cover any desired area and can have its defences enabled or disabled via the master control unit's control panel. Suppose that the house shown in *Figures 4.1* to *4.5* is divided up into the following four defence zones:

Zone 1 = External defences (shed, side-entrance, patio and garden).
Zone 2 = Entire upper floor of the house.

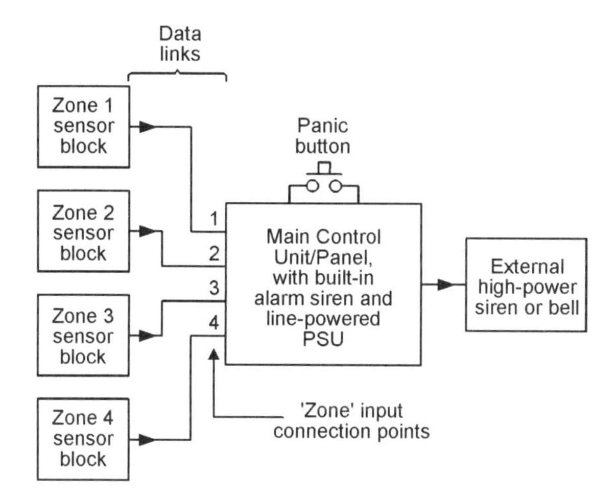

Figure 4.6 *Block diagram of a basic burglar alarm system*

Zone 3 = The ground floor, except the lounge and the hall pressure mat.
Zone 4 = The lounge and the hall pressure mat.

With this defence system, superb round-the-clock anti-burglary protection can be obtained by switching the zones in the following ways, to suit the following circumstances:

When the house is empty, all four zones should be enabled, thus giving total protection. When the house is occupied during normal daylight hours but the garden area is unused, only Zone 1 should be enabled. If only the garden and ground floor are in use, only Zone 2 should be enabled. In the evening, if only the lounge is in use, Zones 1, 2 and 3 should all be enabled, thus protecting the occupier against the opportunist burglar who sneaks into the building while the family is watching TV in the evening (almost a quarter of all domestic burglaries occur when the house is occupied, with the occupiers either watching TV or asleep in bed). At night, when only the upper floor of the house is occupied, Zones 1, 3 and 4 should all be enabled. Note that the 'panic' button is normally enabled even when all four zones are disabled, thus giving the owner non-stop protection against thugs.

Entry/exit delay

In the basic type of system described above, the Zone 1, 2 and 4 defence circuitry is arranged so that, when these zones are active, the alarm sounds instantly if an intrusion is detected, but the Zone 3 defence circuitry (which protects the front door entry/exit and control panel areas) has a built-in 'entry/exit' operating delay of (typically) about 45 seconds. This delay gives the system's key-holder limited freedom to pass through the Zone 3 defence area (to enter or leave the house or to operate the control panel) without sounding the alarm. This delay action is such that, when the owner enables Zone 3 via a key-switch prior to leaving the house or going to bed, the zone does not become active until the end of the 45 second 'exit delay' period; when the owner later passes through Zone 3 again to deactivate the zone via the key-switch, the zone's sensors instantly detect the intrusion and activate a low-level 'warning' bleeper, but only activate the main alarm siren if the owner fails to disable the zone (or reset the alarm) via the key-switch by the end of the 45 second 'entry' delay period.

Thus, the Zone 3 entry/exit delay facility allows the alarm's key holder to move reasonably freely about the house without activating the main alarm siren, but gives full protection against unwanted intruders who do not have access to the security key.

The main control unit/panel

The main control unit is the effective 'heart' of the burglar alarm system, and can be managed via a control panel. In simple units, the panel enables the unit's main functions to be selected via a 4-way master key-switch (usually marked TEST, OFF, ON, and RESET). Most modern control units take the basic form shown in *Figure 4.7* and are powered from the domestic AC power lines via a built-in DC PSU that also provides an auxiliary DC power output. The control unit should ideally also have a built-in rechargeable battery that is normally trickle-charged by the PSU but takes over the PSU's main functions if the AC supply fails or is deliberately interrupted.

Modern control units usually have a built-in medium-power siren, plus a facility for activating a high-power external siren; ideally, the external siren should (to minimize the chances of generating publicly annoying false alarms) not activate until at least 30 seconds after the built-in siren has activated, and must (to conform to local noise control regulations) turn off automatically after a maximum period of about 15 minutes. Most units also have a built-in 'tamper' switch that (except when the master key is set to the TEST or RESET positions) activates the built-in self-latching siren if the unit's case is opened.

The unit's control panel usually takes the basic form shown in *Figure 4.7*, but in practice often uses electronic key-pad (rather than electromechanical) control switching. In this diagram, key-operated switch S1 selects the units main functions, and toggle switches S2 to S5 allow individual defence zones to be enabled or disabled. When S1 is set to the TEST position, the unit's tamper switch and the external alarm are disabled, and the internal alarm

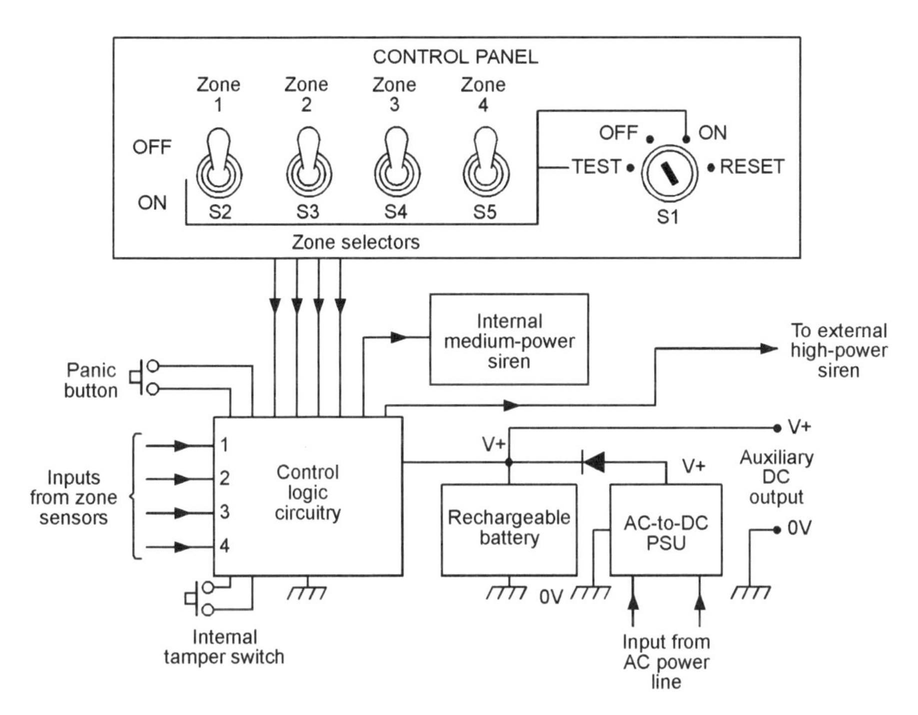

Figure 4.7 *Typical basic form of a modern burglar alarm control unit and its control panel*

operates in the non-latching mode. When S1 is set to the ON positions, the system is fully active and the internal alarm operates in the self-latching mode. When S1 is set to OFF, all four sensor zones are disabled, but the self-latching PANIC facility is fully active. If the alarm is activated in the self-latching mode, the alarm can only be turned off by first removing the cause of activation and then unlatching the alarm by moving S1 to the RESET position.

In traditional hard-wired alarm systems, the sensors that connect to the alarm's various 'zone' input points usually take the effective forms of contact switches, which can easily be enabled or disabled by the S2 to S5 toggle switches shown in *Figure 4.7*. *Figure 4.8* shows the connections for turning individual sections of the alarm sensor network on or off. Series-connected n.c. sensor networks can be enabled or disabled by wiring them in parallel with S2, as shown in *Figure 4.8(a)*; the sensors are enabled when S2 is open, and are disabled when S2 is closed. Parallel-connected n.o. sensor networks can be enabled and disabled by wiring them in series with S2, as shown in *Figure 4.8(b)*; the sensors are enabled when S1 is closed, and are disabled when S1 is open.

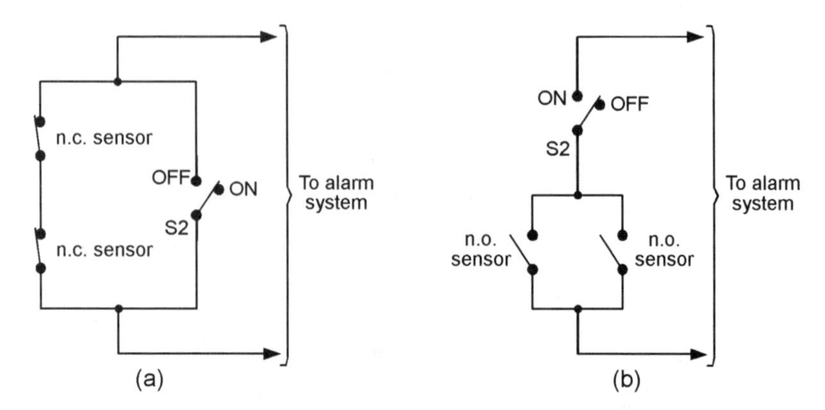

Figure 4.8 *Method of enabling and disabling sensor switches via S2; (a) series-connected, n.c., (b) parallel-connected, n.o.*

The external siren/alarm

The medium-power siren built into most modern control units usually drives an efficient piezoelectric output transducer and operates at a fairly high audio frequency (typically 1.5kHz to 4kHz); this type of siren floods the house with sound, but such sound attenuates rapidly with distance. Sirens designed for external use normally use an efficient horn-type loudspeaker as their output transducer and typically operate in the 800Hz to 1.2kHz audio range, which offers good long-distance acoustic coverage. Both types of siren normally generate an attention-grabbing multi-tone (pulsed, warbled, or swept) sound, rather than a tiresome monotone sound.

External sirens are usually enclosed in a weather-proof bell-type alarm box that is screwed to the house front; the box often incorporates a flashing alarm beacon that activates at the same time as the siren. Low-cost units of this type are usually powered, via a multi-cored cable, from the auxiliary power output terminals of the main control unit, and can be disabled by simply severing the power cable, which must thus be protected by burying it in brickwork, etc. High-quality external siren units, on the other hand, are self-powered and tamper-proof, and should (to conform to current design standards) meet the following basic design specifications:

(1) The siren must be powered by an internal rechargeable battery that is automatically trickle charged in some way and has enough capacity to provide at least 4 hours of continuous alarm operation.
(2) The unit must be designed so that the alarm is not triggered by a temporary failure in the trickle charging system, but will trigger if activated by the main control unit or if the unit's main feed cable is cut.

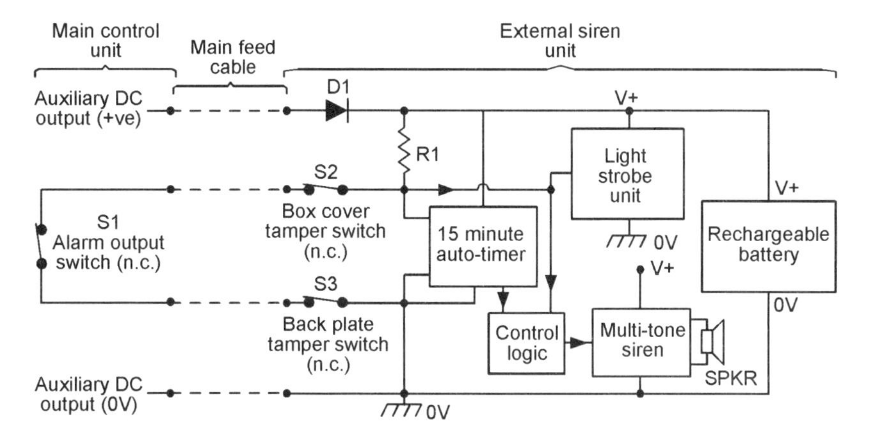

Figure 4.9 *Block diagram of a high-quality external siren unit*

(3) The box must incorporate tamper switches that automatically activate the alarm if the box's cover is removed, or if the complete unit is forcibly removed from its fixing point (the wall).
(4) The unit must incorporate a timing mechanism that automatically resets the siren (but not necessarily the beacon) after (typically) not more that 15 minutes of continuous operation.

Figure 4.9 shows the block diagram of a typical high-quality external siren unit that satisfies the above design criteria, and is powered by a built-in battery that is trickle charged via the control unit's auxiliary DC output terminals. This system is designed so that the control unit's 'alarm' output is connected to the siren unit's input via normally-closed (n.c.) switch S1, which is loop-wired in series with the siren unit's two built-in n.c. tamper switches; one of these (S2) is connected to the unit's box cover and opens if the cover is removed; the other (S3) is connected to the unit's back plate and opens if the unit is pulled away from the wall. Thus, the siren activates if S1 opens, or if the main feed cable is cut, or if the box cover is removed, or if the unit's back plate is torn from the wall. If any of these conditions occur, the unit's built-in light strobe unit activates for the duration of the open circuit condition (or until the battery is exhausted), but the multi-tone siren (which is controlled by a 15-minute auto-timer and control logic) activates as soon as the open circuit condition occurs but resets again when the open circuit condition ends or, if the open circuit condition persists, after a maximum of 15 minutes. The siren may then turn off until the control unit is reset, or may reactivate again after a short delay, depending on the auto-timer design.

Note that the unit shown in *Figure 4.9* is connected to the main control unit via a 4-core cable, with two cores dedicated to the trickle charger function and the other two to the 'alarm' signal. In some units, however, the siren and light strobe units are individually activated by the main control unit, and in such cases a 6-core cable may be used.

Wired versus wireless alarm systems

Modern commercial burglar alarm systems are usually microcontroller based, use a key-pad type of control panel, and incorporate an event recorder that – if a break-in occurs – records the precise order in which the various defence zones are invaded. Such systems come in two basic types, being either 'wired' or 'wireless' systems. In wired systems, all zone sensors (PIR movement detectors, contact switches, etc.) are cable wired to the main control unit, which in turn is wired to the external siren unit; such units are time consuming and (since they use lots of interconnection cable) messy to install. In wireless systems, all major zone sensors incorporate a wireless Tx unit that communicates (via a coded 418MHz or 458MHz RF signal) with a matching wireless Rx unit that is built into the main control unit; the Tx unit signals to the Rx unit if an alarm, tamper, or low-battery condition occurs; such systems are very easy and clean to install, but are considerably more expensive that normal wired systems.

All domestic wireless systems are provided with a key-fob style Tx unit that can be used to remotely set or unset the main alarm unit and to act as a 'panic' switch that can remotely activate the alarm siren at any time. Sensor units such as contact and PIR transmitters are battery powered (usually by a PP3 type battery) and, in approved designs, give at least six month of continuous operation per battery charge. Such units are permanently active, can transmit pre-settable identification codes, have a built-in tamper switch that initiates a full alarm condition if the unit is illegally opened, have a low-voltage detector that warns of a failing battery condition, and incorporate sophisticated energy-saving circuitry that greatly extends battery life; PIR units, for example, transmit a brief alarm signal as soon as an intrusion is detected, but then automatically go into a (typically) 60 second shut-down mode before becoming active again; this technique conserves power when a defended area is in normal 'zone off' use, but gives an instant intrusion warning if the zone is alarm-active.

Domestic-type wireless alarm systems vary greatly in price and performance. The cheaper systems usually provide a total of 256 possible identification codes and only four sensor defence zones, to which the sensors can be individually matched by a built-in block of ten miniature switches arranged as shown in *Figure 4.10*. Here, switches 1 to 8 enable the sensor unit's 8-bit Tx 'system' code to be matched to that used by the system's RX unit (which can be pre-set by the owner), and switches 9 and 10 are used to allocate a 2-bit 'zone' number to the sensor. At the other end of the price

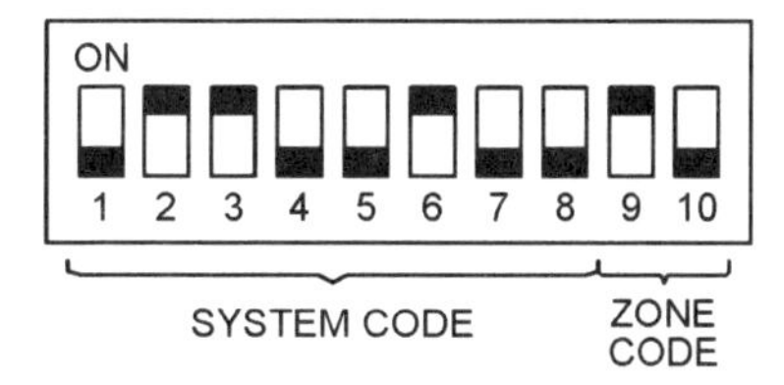

Figure 4.10 *Block of ten miniature switches built into each wireless sensor used in a low-cost wireless burglar alarm, to match the sensor's Tx 'system' and 'zone' coding to that used in the control unit's wireless Rx section*

scale, some systems offer 10- or 12-bit identification codes, and up to 16 defence zones (identified by a 4-bit 'zone' code).

Note that, in wireless systems, the main control unit identifies individual sensors purely by their zone codes. Thus, if a sensor in (say) Zone 3 transmits a low-battery-voltage warning, the unit's control panel will display the fact; if only one sensor carries the Zone 3 code, the owner can quickly identify this particular unit and change its battery, but if several sensors carry the same code the fault can only be traced by individually testing all of the Zone 3 sensor batteries.

Also note that wireless 'contact' sensors usually activate only when an n.c. input switch opens for a period of at least 200mS (this technique minimizing the chances of false alarming due to transient switching or signal pick-up); such sensors can be used with any desired number of n.c. sensor switches (reed-and-magnet switches, window foil, etc.) that are loop wired in the basic manner shown in *Figure 4.11*; they cannot be directly used with pressure mat switches, which are n.o. devices.

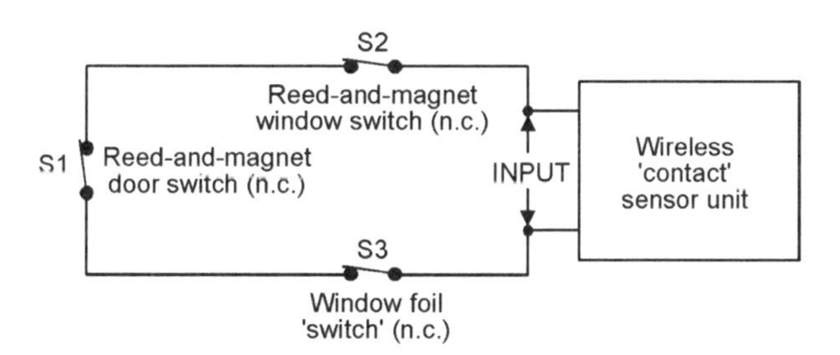

Figure 4.11 *Example of three n.c. sensor switches loop-wired to the input of a single wireless 'contact' sensor unit*

Figure 4.12 shows a simple relay-based adaptor circuit that can be used to activate a wireless contact sensor unit via a pressure mat switch or any other n.o. type of switch (or by any desired number of parallel-connected n.o. switches). Here, 12V relay RLA has a coil resistance of at least 270R, and

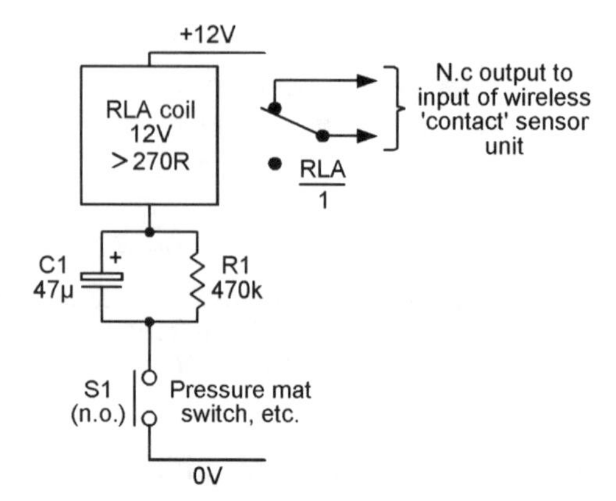

Figure 4.12 *Simple adaptor circuit can be used to activate a wireless contact sensor unit via an n.o. switch such as a pressure mat type*

has one set of change-over (c.o.) contacts that have their n.c. pins wired to the input of the wireless sensor unit. RLA's coil is wired in series with the 12V supply via mat switch S1 and the parallel C1-R1 combination. Normally, S1 is open, C1 is fully discharged, and the RLA/1 output contacts are closed. If S1 now closes, a heavy pulse of current flows through RLA coil via C1 and S1, thus opening the RLA/1 output contact and activating the wireless contact sensor unit. If S1 remains closed, the RLA current rapidly decays to a very low value (determined by R1) and (after a few hundred milliseconds) the RLA/1 contacts re-open; when S1 opens again, C1 slowly discharges via R1 until – after a delay of a minute or so – the system can again be reactivated by closing S1. This circuit thus draws zero quiescent current, and draws only a few microamps of mean current if S1 is closed for long periods.

Wireless alarm system categories

Domestic wireless burglar alarm systems vary greatly in price and performance, but can be roughly divided into the categories of 'low-cost', 'mid-range', and 'top-of-the-range' types. *Figures 4.13* to *4.15* illustrate the basic features of typical examples of each of these system types.

The cheapest and most popular types of wireless burglar alarm system are those that give levels of protection that are quite adequate for use in small flats or apartments, but give only very basic protection when used in 2 or 3 bedroom houses. *Figure 4.13* illustrates, in block diagram form, the basic features of a typical low-cost system of this type. This system offers a total of four defence zones, but these are not individually selectable; in the unit shown, the user has the simple option of making either all four zones active (when the premises are empty), or of making all but one zone (Zone 4, the

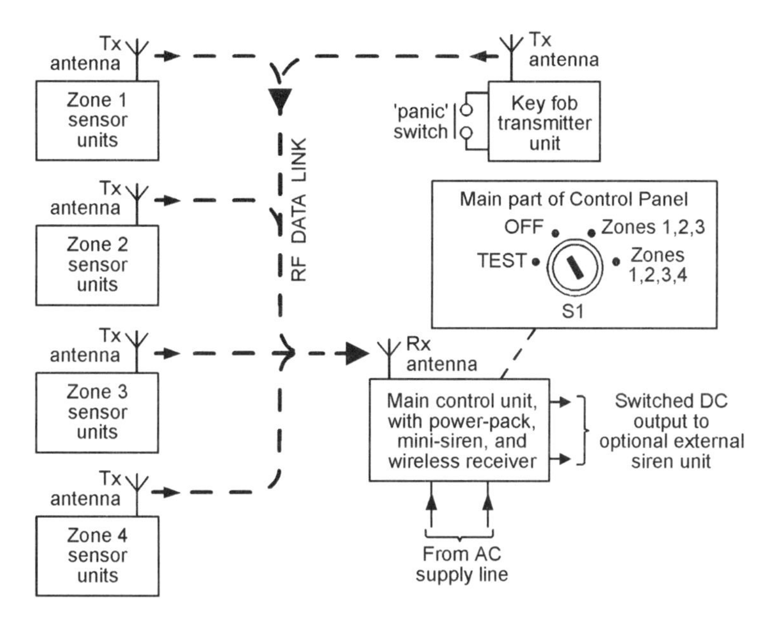

Figure 4.13 *Block diagram showing the basic features of a typical 'low-cost' wireless burglar alarm system*

sleeping and bathroom areas) active (when the occupants are resting); the system gives no protection against burglars who enter the house while the occupants are watching TV, etc. Systems of this type rely on the control unit's built-in siren to scare off any intruders; often, they are not supplied with an external siren but have provision for driving an optional external siren that is powered by the control unit; such sirens can be disabled by simply cutting their feed cables.

Figure 4.14 illustrates the basic features of a typical 'mid-range' wireless burglar alarm system that is designed for use in most houses and in small commercial premises. This system offers a total of six defence zones, all of which are individually selectable, and offers good protection against all types of burglar, including those who enter the house while the occupants are watching TV, etc. Systems of this type are usually supplied complete with an internally-powered external siren/strobe unit that is cable-wired to the main control unit and is fully protected against tampering and cable-cutting, etc.

Finally, *Figure 4.15* illustrates the basic features of a typical 'top-of-the-range' wireless burglar alarm system that is designed for use in large houses or medium-sized commercial premises. This system offers a total of twelve defence zones, all of which are individually selectable, and offers excellent overall protection. The system shown is completely wireless, with no cable link between the main control unit and the external siren/strobe unit, which is wireless-activated (by the control unit), is fully protected against tampering,

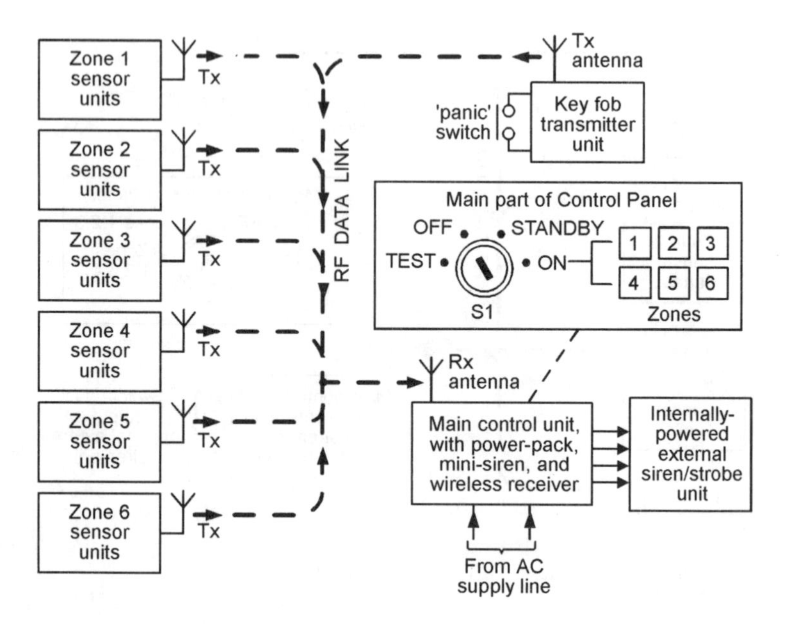

Figure 4.14 *Block diagram showing the basic features of a typical 'mid-range' wireless burglar alarm system*

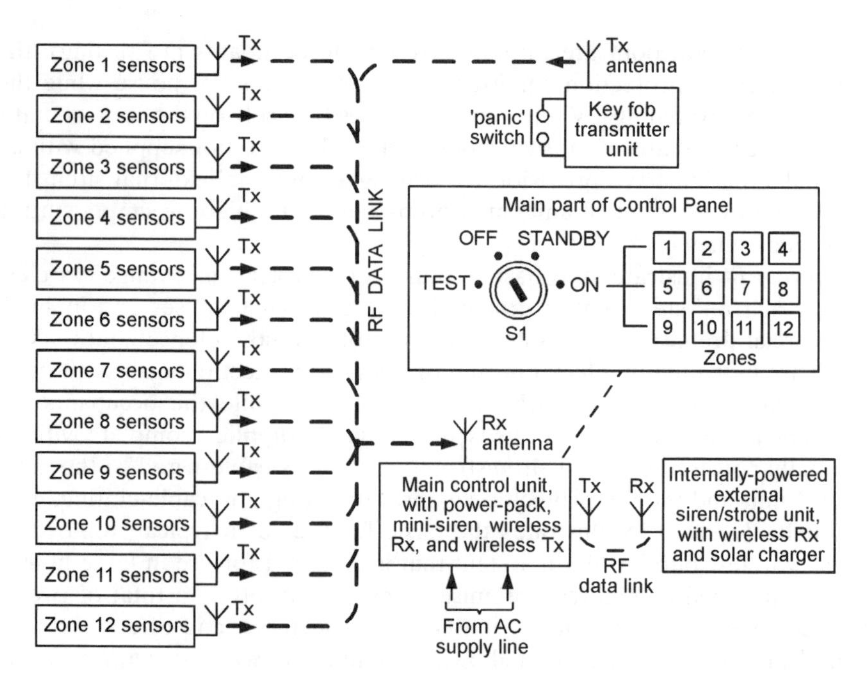

Figure 4.15 *Block diagram showing the basic features of a typical 'top-of-the-range' wireless burglar alarm system*

and is powered by an internal battery that is trickle charged by an integral solar panel.

Intrusion sensor types

The two types of intrusion sensor most widely used in modern domestic burglar alarm systems are PIR movement detectors for 'area' protection, and reed-and-magnet switches for 'spot' protection on doors and windows, etc. Other types of sensor commonly used in domestic systems are pressure mat switches for spot 'floor' protection, vibration sensors to give 'object' protection, and window foil and glass break detectors to detect window breakage. All of these sensors are similarly used in burglar alarm systems designed to protect commercial premises, which sometimes also use IR light beams or brittle wires (built into walls, floors, or ceilings) to detect break-ins via the building's shell.

Note that modern PIR movement detectors are relatively inexpensive and have a high immunity to false alarms, and have consequently replaced once-popular but unreliable capacitive proximity detectors and microwave and ultrasonic movement detectors in most modern commercial 'area' protection systems. Older readers may also note that once-popular 'dual-purpose loop' burglar alarm systems – in which all contact sensors are wired to a continuously-monitored loop that is fitted with an end-of-line resistor – are no longer used in the modern domestic security systems.

Practical burglar alarm circuits

Introduction

Modern microcontroller-based domestic burglar alarm systems are, like TVs and many other electronic 'consumer' products, so reasonably priced that few people would seriously consider DIY-building – rather than buying – such products. This is particularly true of wireless burglar alarm systems, which must use wireless Tx sections that have passed stringent tests laid down by a government testing/licensing authority. It is, however, possible to cost-effectively DIY-build a variety of fairly simple and inexpensive 'conventional' burglar alarm and accessory circuits, and a number of these are described in the remaining sections of this chapter.

A 'false key' booby trap circuit

Some people hide a spare front door key under a flower pot or porch mat or on a porch ledge when they leave the house, and burglars often make a

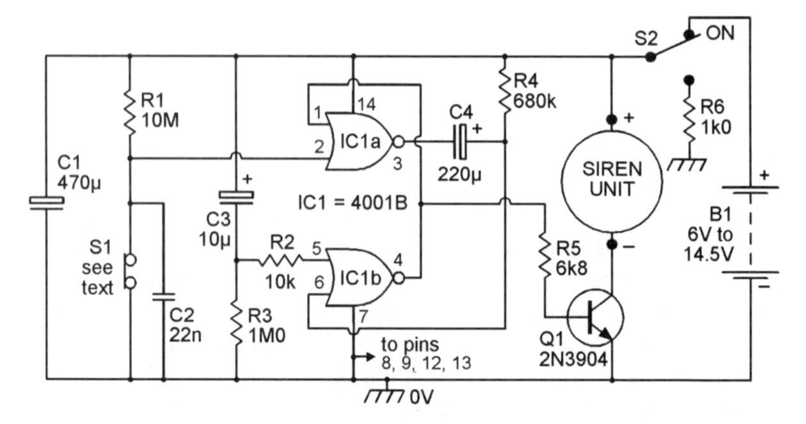

Figure 4.16 *'False key' booby trap alarm circuit*

quick search for such a key when they enter the porch area of a house. 'False key' booby trap units take advantage of this fact by activating a semi-latching siren if an object such as a flower pot is briefly moved, or if someone grabs a key that is tied to a short length of string, etc. *Figure 4.16* shows the practical circuit of such a unit.

In *Figure 4.16*, two gates of the 4001B CMOS IC are wired as a simple monostable multivibrator that gives a low pin-4 output when a positive voltage is fed to pin-5, but, if pin-5 is low, produces a positive pin-4 output pulse if a positive-going transition is applied to pin-2 by opening sensor switch S1; this output pulse has a duration of about 2 minutes with the R4–C4 values shown, and is used to activate an inexpensive commercial siren unit via R5 and transistor Q1. Note that the monostable can only be triggered by a positive-going transition of its pin-2 voltage; its action is not influenced by 'standing' high or low voltages applied to pin-2 via R1–S1. Thus, the action of this booby trap circuit – which may typically be housed in an inverted flower pot – is as follows:

When power is switched to the circuit via S2, C3–R3 apply a decaying positive voltage to pin-5 that disables the monostable for at least 12 seconds, thus giving the user time to safely 'prime' the circuit (position it so that S1 is held in the closed position) without activating the siren. At the end of this period the monostable becomes enabled; if sensor switch S1 subsequently opens for a period in excess of 200mS (determined by R1–C2), the monostable fires and activates the siren for a continuous period of about 2 minutes (which is long enough to scare off most burglars). At the end of this 2-minute period the siren turns off, irrespective of the state of S1, and can only be retriggered by closing and then opening S1 again. Note that R6 discharges the circuit's timing capacitors when control switch S2 is turned off, that most of the unit's circuitry must be weatherproofed and protected with varnish, and that the unit can be powered by any 6V to 14.5V battery supply and the unit consumes a quiescent current of only a few μA (mainly via R1 and via C1's leakage currents).

The circuit's S1 sensor switch can take various forms; in a flower pot unit it may be an n.o. key-pad switch that is normally held closed by the weight of the pot, but opens when the pot is lifted; in another case it may be an n.c. type that opens when someone tugs on a piece of string, and so on.

Shed/garage burglar alarm circuits

Domestic workshops and garages that are fitted with AC power lines are best defended by simple AC-powered 'house/flat' types of burglar alarm that can activate powerful siren/light-strobe units for several minutes, and a versatile alarm unit of this type is shown later in this chapter. Most garden sheds and many domestic garages (and also caravans and small boats, etc.) are, however, devoid of AC power lines, and are best defended by battery-powered burglar alarms that, when activated, sound a siren for only a few minutes; this section looks at some practical circuits of this type.

Figures 4.17 to *4.20* show alternative versions of battery-powered shed/garage burglar alarms, which should ideally be powered by rechargeable batteries that are kept fully energized by solar-powered charger units. The *Figure 4.17 – 4.18* unit is meant to be turned on and off by a key-switch that is operated from outside the building; the *Figure 4.19 – 4.20* unit is meant to be turned on and off from within the building, and incorporates exit/entry

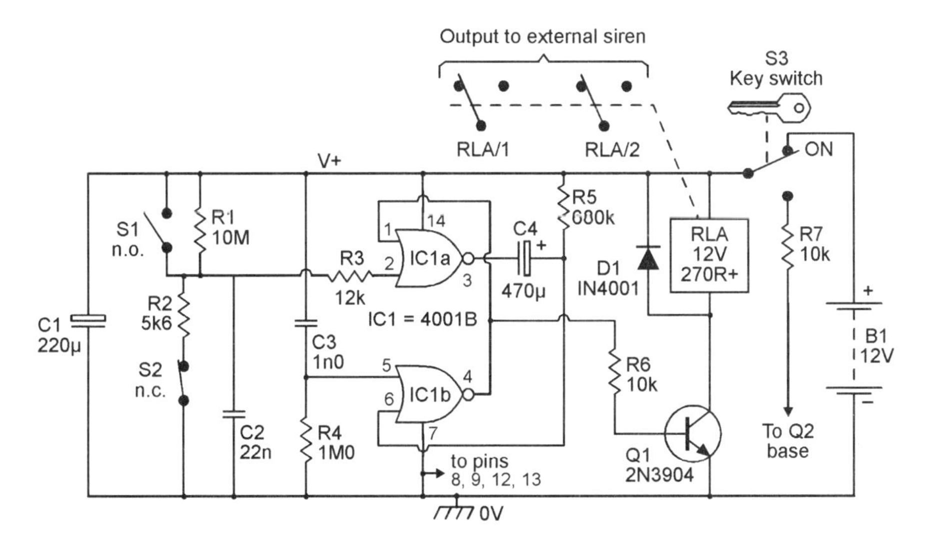

Figure 4.17 *Shed/garage burglar alarm, with external on/off switch*

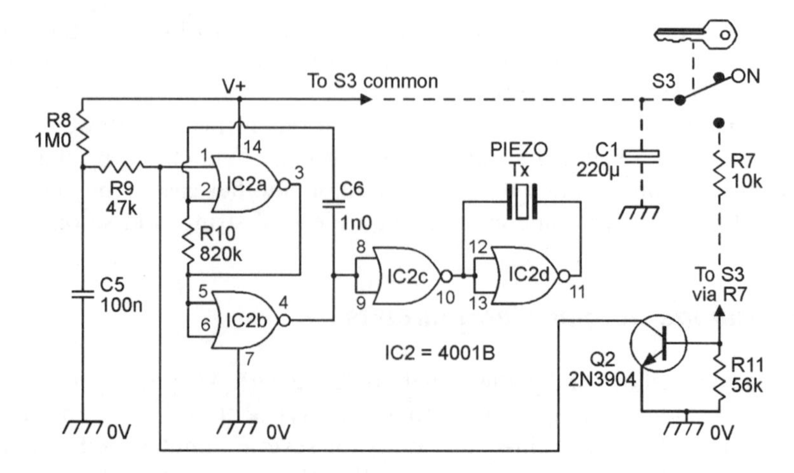

Figure 4.18 *Optional state-indicating sounder, for use with the Figure 4.17 circuit*

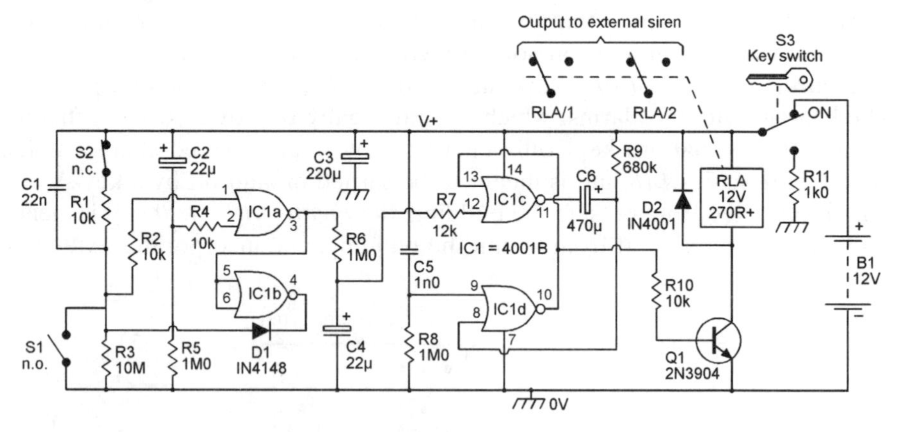

Figure 4.19 *Shed/garage burglar alarm, with internal on/off switch*

time delays that let the key holder leave and enter the building without sounding its alarms. Each burglar alarm unit consumes a typical ON ('standby') current of 1-2µA, can use any desired number of n.o. (S1) and/or n.c. (S2) sensor switches, and has a pair of c.o. relay output contacts that latch on for about 5 minutes under the 'alarm' condition and can be used to activate any type of external siren, which may be self-powered or may be powered from the burglar alarm's battery via the relay contacts.

Note in the *Figure 4.17* and *4.19* diagrams that S1 can consist of any desired number of n.o. switches (including 'tilt' switches of the type used on up-and-over types of garage door) wired in parallel, and S2 can consist of any desired number of n.c. switches (such as reed-and-magnet switches on shed doors

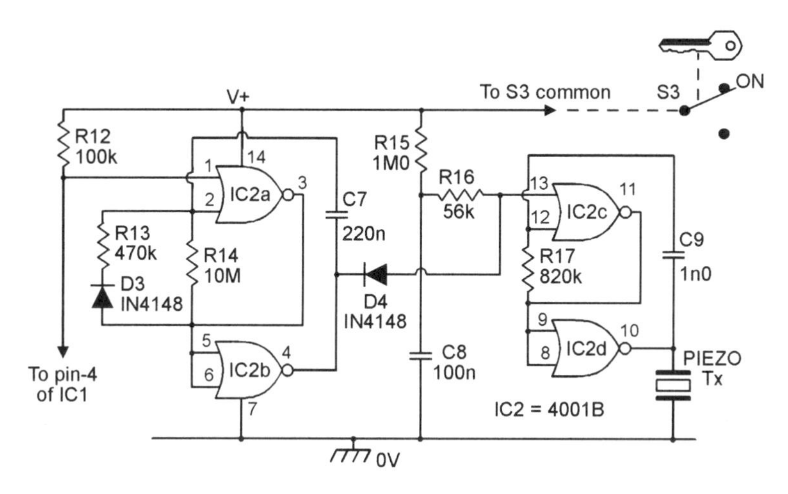

Figure 4.20 *State-indicating sounder, for use with the Figure 4.19 circuit*

and opening windows, anti-tamper switches built into alarm and siren boxes, wire 'loops' formed inside easily-cut cables, and cable loops used to protect tools, etc.) all wired in series; if S1 is not needed, simply omit it; if S2 is not needed, replace it with a short.

The *Figure 4.17* burglar alarm circuit is turned on and off via key-operated switch S3, which is mounted in a position where it can be operated from *outside* of the building's main entrance; thus, S3 is used to enable the alarm after leaving the building, and to disable it before entering the building. The circuit is basically similar to that of *Figure 4.16*, which uses two gates of a 4001B IC as a triggered monostable pulse generator. In this case, however, the monostable output has a period of about 5 minutes and activates relay RLA via transistor Q1, and can be triggered by closing n.o. switch S1 or by opening n.c. switch S2.

Figure 4.18 shows an optional audible-output 'system-state' indicator that can be added to the *Figure 4.17* alarm circuit and emits a brief 'bleep' when the alarm circuit is first switched on, confirming that it is receiving power, and emits a longer 'decaying' bleep as the alarm circuit is switched off, confirming that its power has been removed (if you do not use this add-on circuit, change the R7 value to 1k0 and wire its low end to the 0V line). In *Figure 4.18*, IC2 is wired as a gated astable that – when gated on by a 'low' voltage on pin-1 – generates an audible tone signal in a low-cost piezo sounder. The action is such that the astable is briefly driven on via R8–C5 as S3 is switched to the ON position, thus generating a brief 'bleep' in the sounder; when S3 is switched to the OFF position, C1's stored charge drives the astable on via R7–Q1 and supplies the astable with limited operating power, thus producing a decaying 'bleep' in the sounder.

The *Figure 4.19* burglar alarm circuit is turned on and off via key-operated switch S3, which is mounted *inside* the shed/garage; when S3 is first turned ON, an 'exit delay' comes into operation, giving the key holder about 18 seconds to leave the building, after which all S1/S2 sensor switches become fully active; when the building is re-entered after this period, the sensor switches trigger an 'entry delay' timer that – if S3 is not switched OFF within 18 seconds – triggers a 5-minute monostable that drives an external siren via the contacts of relay RLA. The basic circuit is similar to that of *Figure 4.17*, except that exit/entry time-delay logic is interposed between the outputs of S1/S2 and the input trigger point of the 5-minute monostable pulse generator (IC1c–IC1d). The circuit operates as follows:

In *Figure 4.19*, IC1a is used as a NOR gate that gives a low (logic-0) pin-3 output if either input is high, and gives a high output only if both inputs are low. The pin-1 input of IC1a is normally high, but goes low if S1 closes or S2 opens; the pin-2 input of IC1a is normally low, but is held high by the C2–R5 'exit delay' network for about 18 seconds when power is first applied to the circuit via S3. Thus, IC1a's output is locked low during the 'exit delay' period, but can subsequently switch high if S1 closes or S2 opens; if this latter action occurs, the output of inverter IC1b pulls IC1a's pin-1 input low via D1, thus locking its output into the high state, irrespective of subsequent S1/S2 actions. This 'high' output is fed to the pin-12 'trigger' input pin of the IC1c–IC1d relay-driving 5-minute monostable via the R6–C4 'entry delay' timing network, which triggers the monostable about 18 seconds after pin-3 goes high.

Figure 4.20 shows an optional audible-output 'system-state' indicator that can be added to the *Figure 4.19* alarm circuit and emits a brief 'bleep' when the alarm circuit is first switched on prior to leaving the building, and emits a series of 50mS 'bleeps' at roughly 1-second intervals when anyone re-enters the building, reminding them to turn S3 OFF before the siren-activating finish of the 'entry delay' period. In *Figure 4.20*, IC2c–IC2d are wired as a gated astable that – when gated on by a 'low' voltage on pin-13 – generates an audible tone signal in a low-cost piezo sounder, and IC2a–IC2b are wired as a gated asymmetrical astable that gates the IC2c–IC2d astable via D4 and activates automatically when anyone re-enters the building. The action is such that the IC2c–IC2d astable is briefly driven on via R15–C8 as S3 is switched to the ON position, thus generating a brief 'bleep' in the sounder, and is activated via the IC2a–IC2b astable whenever the main unit's 'entry delay' circuitry becomes active, thus generating a series of 50mS 'bleeps' that are repeated at 1-second intervals until the main alarm unit is turned off via S3.

House/flat burglar alarm circuits

Shed/garage burglar alarms of the *Figure 4.17* to *4.20* types are simple battery-powered single-zone units. Modern burglar alarms suitable for use in

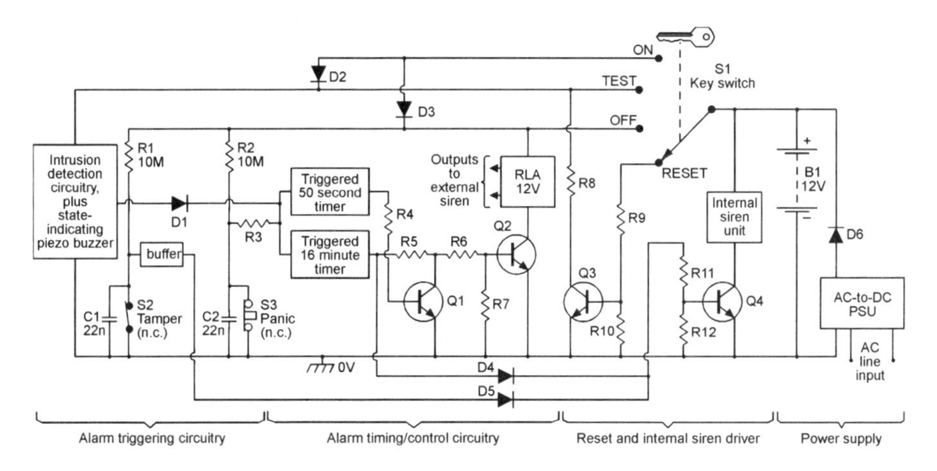

Figure 4.21 *Block diagram of the 'universal' burglar alarm system*

houses, flats and apartments are moderately complex AC-powered multi-zone units with built-in 'panic' and 'tamper' facilities; they usually have an internal trickle-charged battery that provides power in the event of an AC power-line failure, and have auxiliary 12V DC outputs suitable for powering external PIR movement detectors, etc. *Figures 4.21* to *4.25* show the block diagram and practical circuit details of a sophisticated modular 'universal' burglar alarm unit of the latter type that can easily be built to suit the precise needs of the individual user.

Figure 4.21 shows the basic block diagram of the 'universal' burglar alarm unit, which can be fitted with one exit/entry zone plus any desired number of 'normal' defence zones, all of which are individually switch-selectable; each zone is provided with its own audio/visual 'state' indicator (not shown in this diagram) and activates internal and external alarm sirens when an intrusion is detected. The unit can also be fitted with any desired number of n.c. 'panic' switches (S3) wired in series, and with any number of n.c. tamper switches or loops (S2) wired in series. The unit comprises the four major sections shown in the diagram, and offers the following modes of operation, which are selectable via 4-way key-operated switch S1:

ON. When S1 is in the ON position, all four major sections of the unit are energized, and the 'alarm timing/control' circuitry's 50-second and 16-minute timers are immediately triggered if an intrusion is detected by the 'alarm triggering' circuitry or if PANIC switch S3 is opened for more than 200mS. Under this condition the internal siren is immediately driven on via D4 and Q4, but the external siren (which is activated via Q2 and RLA) is held off for 50-seconds via R4–Q1, thus minimizing the chances of *accidentally* sounding the external alarm; both alarms switch off when S1 is moved to the

RESET position, or turn off automatically at the end of the 16-minute timing period; the intrusion detector's piezo buzzer also sounds if an intrusion is detected and operates for the duration of the intrusion condition. The internal siren is driven on (via D5) if TAMPER switch S2 opens, and sounds for the duration of the open circuit condition.

TEST. When S1 is in the TEST position, the 'alarm timing/control' circuitry and the TAMPER and PANIC switches are disabled, but the intrusion detection circuitry is fully active; if an intrusion state is detected under this condition, only the internal piezo buzzer is activated. This mode is useful when testing or checking sensor switches, PIR units, or sensor wiring, etc.

OFF. When S1 is in the OFF position, the intrusion detection circuitry is disabled, but the TAMPER and PANIC circuitry is fully active. If TAMPER switch S2 opens, the internal siren is driven on (via D5–Q4) for the duration of the open circuit condition. If PANIC switch S3 opens, the internal siren is driven on immediately (via D4–Q4) and the external siren activates 50-seconds later; both sirens turn off when S1 is moved to RESET, or turn off automatically at the end of the 16-minute timing period.

RESET. When S1 is in the RESET position, the entire circuit (except the power supply circuitry) is effectively disabled, and Q3 rapidly resets the intrusion detector circuit's 'exit delay' timer.

Figure 4.22 shows the basic circuitry of the 'universal' alarm unit's intrusion sensing/signal processing circuitry, specifically applied to a unit with one entry/exit zone (Zone 'A') and three 'normal' zones (Zones 'B' to 'D'); additional 'normal' zones can be added by simply duplicating the Zone 'D' and D5 circuitry for each extra zone. All zones use the same intrusion sensing circuit design as shown for Zone 'A'. Each zone can use any desired number of series-connected n.c. (SWa) and/or parallel-connected n.o. (SWb) sensor switches, and is selected by a DPDT switch (SWc) that, when closed, connects the output of inverting buffer IC1a to a state-indicating LED (LED1a) and also connects the +12V supply to any auxiliary units (PIRs, etc.) that are associated with the zone. When a zone is selected by SWc and key-switch S1, its action is such that the output of IC1a goes high and illuminates the LED and activates a piezo buzzer (via D1 or D2) if any of the zone's intrusion-detecting sensor switches are activated; this 'high' signal is also passed through the unit's signal processing circuitry, as described in the next two paragraphs.

In *Figure 4.22*, the output signals from the sensing circuits of all selected 'normal' defence zones are ORed via D3–D4–D5 and are then fed to input A (monotone sound) of the piezo buzzer via D2, and also through transient suppressor R10–C3 (which only passes signals that switch high for at least 200mS); the output of R10–C3 is then inverted by IC3a and passed to one

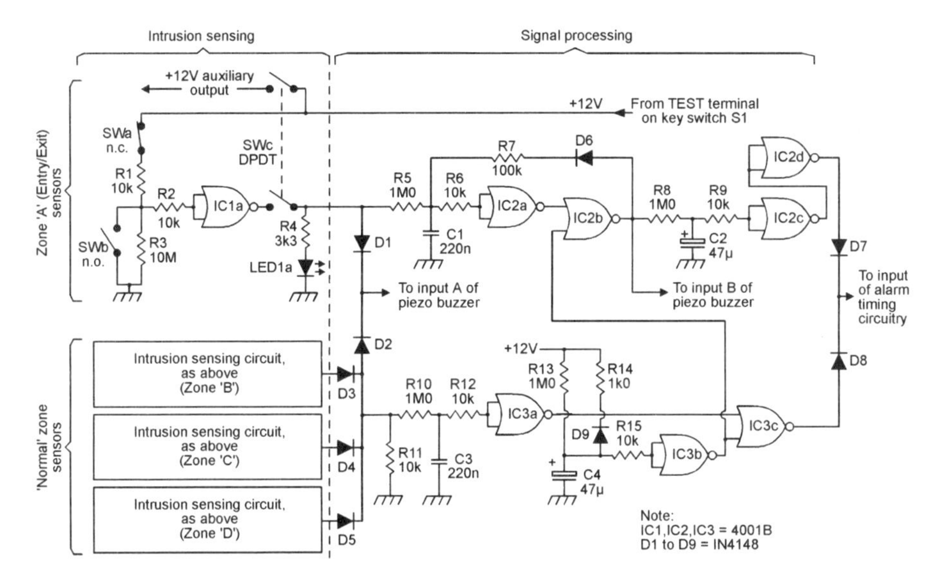

Figure 4.22 *Basic circuit diagram of the intrusion sensing/signal processing circuitry*

input of NOR-gate IC3c, which has its other input derived from 30-second switch-on delay generator R13–C4–IC3b, which disables IC3c for 30 seconds when power is first connected to the circuit via S1. The net result is that the circuit gives an instant audio-visual indication if the output of any selected zone switches high, but under this condition IC3b's output only generates a siren-activating signal (via D8) if the circuitry has been energized via S1 for at least 30-seconds.

In *Figure 4.22*, the output signals from the entry/exit zone are (when SWc is closed) fed to input A of the piezo buzzer via D1, and also passed – via transient suppressor R5–C1 – to the input of a gated self-latching non-inverting buffer formed by IC2a–IC2b, which is gated by the R13–C4–IC3b 30-second switch-on delay generator, which provides the zone with its 'exit' delay. If the zone's output switches high *during* the 30-second exit delay period (as, for example, when someone exits the zone), the circuit gives an instant audio-visual indication of the fact but produces no other effects. If the zone's output switches high *after* the end of the 30-second exit delay period (as, for example, when someone re-enters the zone), the circuit again gives an instant audio-visual indication of the fact, but in this case the output of IC2b switches high and is latched into that state via D6–R7; this action drives input B (timing-beat sound) of the piezo buzzer high and also initiates a 30-second entry delay timing period (controlled via R8–C2 and non-inverting buffer IC2c–IC2d); if the complete circuit is not switched off (via

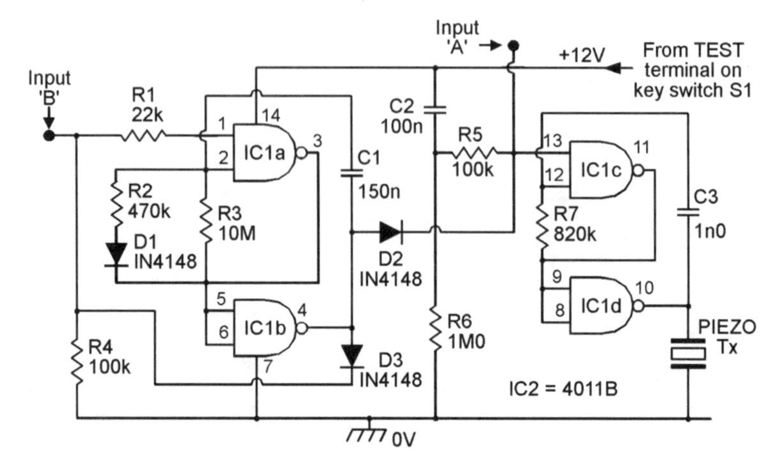

Figure 4.23 *Circuit of the 'universal' burglar alarm's state-indicating piezo buzzer unit*

S1) by the end of this 30-second 'entry' period, IC2d's output switches high and activates the unit's sirens via D7.

Note in *Figure 4.22* that D9–R14 are used to rapidly discharge C4 via the positive supply rail and *Figure 4.21*'s transistor Q3 when S1 is moved to the RESET position. Also note, when building the *Figure 4.22* circuit, that pin-14 of all 4001B ICs must be wired to the +ve supply rail, pin-7 to the 0V rail, and that all unused gate input pins must be tied to the 0V rail. All LEDs must be high-brightness types.

Figure 4.23 shows the circuit of the unit's state-indicating piezo buzzer unit, which is powered from the supply rails of the intrusion sensing/signal processing circuitry and consists of two gated astables that are activated by high (logic-1) gate voltages. IC1c–IC1d are wired as a simple 'tone' astable that – when gated on via pin-13 – generates a 680Hz tone signal in the piezo sounder, and IC1a–IC1b are wired as a gated semi-latching asymmetrical astable that – when gated on via the alarm's 'entry delay' timer (see *Figure 4.22*) – produces one-per-second 50mS output pulses that gate the tone astable via D2. The action is such that the tone astable is briefly driven on via C2–R6 when power is first switched to the circuit, thus generating a brief 'switch-on' bleep in the sounder, and is activated via the input-'A' terminal whenever a sensor switch is activated in any of the alarm's active zone areas. The tone astable is also activated via the IC1a–IC1b astable and D2 whenever the alarm's 'entry delay' circuitry becomes active, thus generating a series of 50mS 'bleeps' at roughly 1-second intervals when anyone re-enters the building, reminding them to turn key-switch S1 OFF before the siren-activating finish of the 'entry delay' period.

Figure 4.24 shows the circuit of the alarm's siren control unit, which is based on the block diagram of *Figure 4.21* but uses its own component

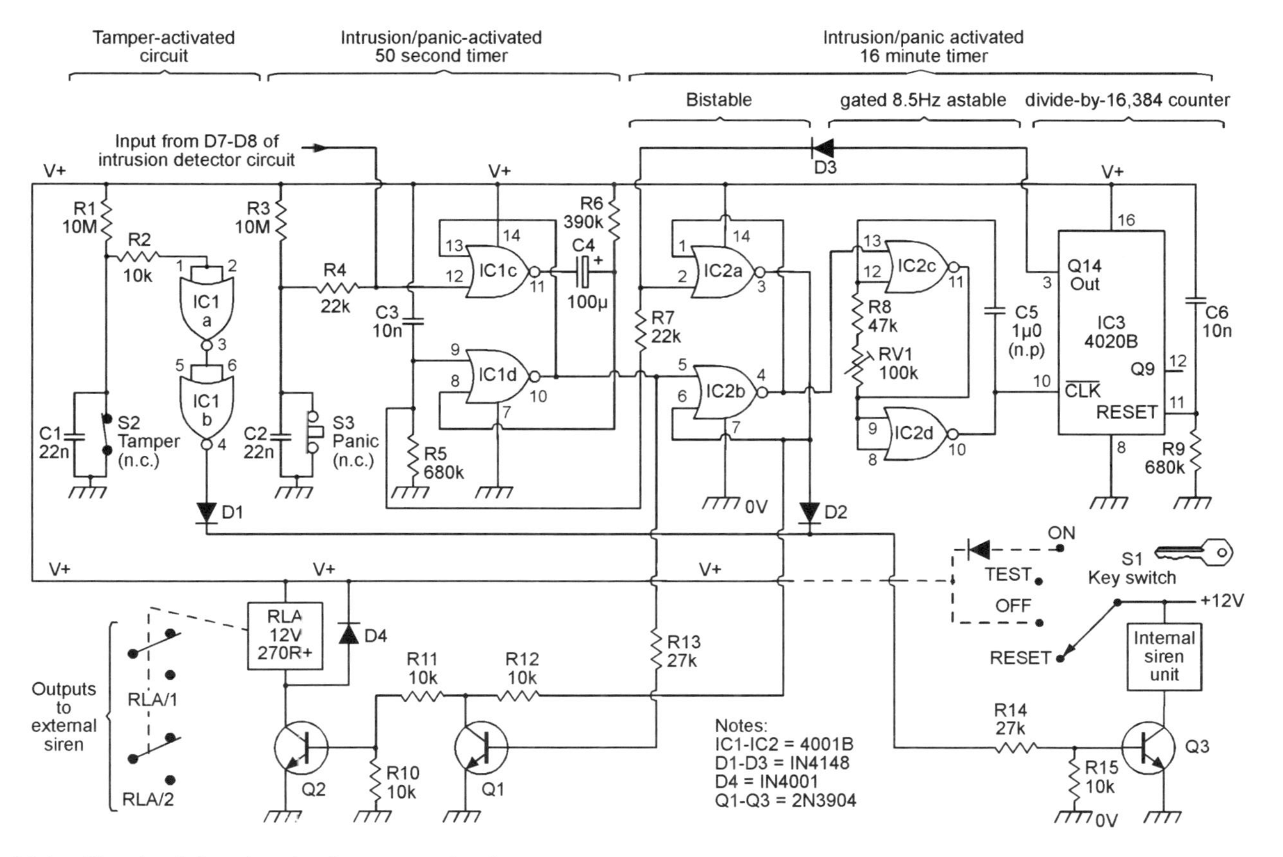

Figure 4.24 *Circuit of the alarm's siren control unit*

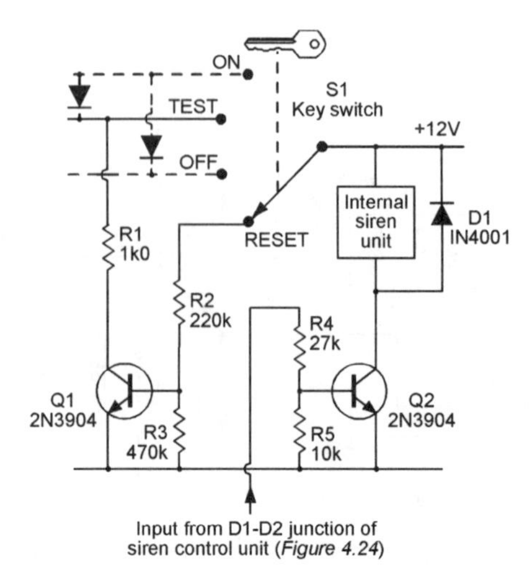

Figure 4.25 *The alarm's 'reset' and internal driver circuitry*

numbering system and is energized when S1 is in the ON and OFF positions. Here, IC1a–IC1b are wired as a non-inverting buffer that drives the internal siren on (via D1–Q3) if Tamper switch S2 opens, and the remaining ICs act as triggered 50-second and 16-minute timers that activate the internal and external sirens if Panic switch S3 is opened or if a 'high' input is received from the output of the alarm's intrusion detector circuitry. These timing circuits operate as follows:

In *Figure 4.24*, IC1c–IC1d are wired as a simple monostable timer that controls the external siren's 'hold-off' period; it is automatically reset at S1-switch-on via C3–R5 and is triggered by a positive-going transition on pin-12 (derived from the intrusion detector, or by opening S3). When triggered, the monostable's pin-10 output switches high and activates the IC2–IC3 16-minute timer and turns Q1 on, but switches low again at the end of its 50-second (nominal) timing period, which is controlled by R6–C4. In practice, this timing period also depends on the 'threshold' voltage value of the individual IC, and may vary substantially from the 50-second value; if it does, make the timing roughly correct by changing the R6 value.

The output of the 50-second timer triggers the 16-minute timer, which is a semi-precision design built around a bistable latch (IC2a–IC2b), a gated 8.5Hz astable (IC2c–IC2d), and a 14-stage (divide-by-16 384) ripple counter (IC3). The action is such that, at switch-on, the bistable is automatically reset (with its pin-3 output low and pin-4 high) via C3–R5–R7, thus gating the astable off, and the counter is reset via C6–R9. As soon as the 50-second timer (IC1c–IC1d) is triggered, its pin-10 output flips the IC2a–IC2b bistable, driving the internal siren on via D2–Q3 and feeding a drive current towards the base of relay-driving transistor Q2 via R12–R11, and also gating on the astable, which immediately starts feeding clock pulses into the IC3 counter at a 8.5Hz rate.

Note that, in the early stages of this 16-minute timing sequence, Q1 is driven on by the monostable timer, thus preventing the bistable's drive current from reaching the base of Q2, but that Q1 turns off after 50 seconds, thus enabling Q2 to turn on and activate the external siren via relay contacts RLA/1–RLA/2. Meanwhile, the astable keeps feeding clock pulses into the counter until, after 16-minutes, on the arrival of the 8192nd pulse, the pin-3 output flips high and resets the bistable via D3, thus terminating the timing process and turning both the internal and external sirens off. The circuits timing period can easily be set to precisely 16-minutes by connecting a LED and 4k7 series resistor between pins 12 and 8 of IC3 and – with the timer triggered – carefully trimming RV1 so that the LED operates with precise 30-second on and off periods.

Figure 4.25 shows the circuit of the alarm's 'reset' and internal siren driver circuitry, together with its connections to S1; this diagram is based on those of *Figures 4.21* and *4.24* but uses its own component numbering system. Here, Q1 is driven on whenever S1 is in the RESET position, and rapidly resets the alarm's intrusion detector 'exit delay' timer by discharging its timing capacitor (C4 in *Figure 4.22*). Q2 is driven on (via the output of the alarm timing/control unit) and activates the internal siren unit (a low-cost multi-tone medium-power commercial unit) whenever an intrusion is detected or a tamper or panic switch is operated; the specified Q2 transistor has a maximum current rating of 200mA, which is adequate for driving most medium-power 12V sirens.

Regarding the 'universal' burglar alarm's power supply, note that the basic unit consumes a typical standby current of only a few microamps and can, if desired, simply be powered by a rechargeable 12V battery. In practice, however, modern burglar alarms are usually used in conjunction with PIR detector units, each of which typically consume a quiescent current of 20mA; thus, a system that uses three PIRs consumes a quiescent operating current of about 60mA, which can – if desired – be supplied by a 12V rechargeable 1.2AH battery that is fed – via a protective diode – via the output current of a line-powered 60mA trickle charger. In a unit of this type, the charger supplies the full operating current when the alarm is in its ON but untriggered mode; the battery supplies all excess power if the alarm is triggered, and receives a safe 60mA (1/20th of its 1.2AH capacity) trickle charge when the alarm is not in the ON state.

Finally, note that the 'universal' burglar alarm is very simple to operate, and is normally used in the ON mode when required to respond to an intrusion, and in the OFF mode (in which its panic and tamper switches are still active) when it is not required to detect an intrusion. The TEST mode is only used when setting up or testing the system. The RESET mode is only used to reset the alarm timing/control circuitry once an alarm siren has activated, or to rapidly reset the intrusion detector's exit delay timer when an unexpected repeat of the full 'exit delay' time is needed.

5

Temperature-sensitive security circuits

Temperature-sensitive security circuits can be used to automatically activate alarms or safety devices when one or more monitored temperatures goes above or below a pre-set level, or when two temperatures differ by more than a prescribed amount. Such circuits can be used to give warning of fire, frost, excessive boiler temperature, the failure of a heating system, or overheating of a piece of machinery or a liquid, etc. They may use thermostats, thermistors or various types of solid-state device as their temperature-sensing elements, and may give an audible output via some type of electro-acoustic device (a siren or bell, etc.) or a switched output via some type of relay. A wide range of useful temperature-sensitive security circuits are presented in this chapter.

Thermostat fire-alarm circuits

One of the simplest types of temperature-sensitive circuit is the thermostat-activated fire alarm. *Figure 5.1* shows the practical circuit of a relay-aided non-latching alarm of this type. Here, any desired number of n.o. thermostats are wired in parallel and then connected in series with the coil of a relay, and one set of the relay's n.o. contacts are wired in series with the alarm bell so that the bell operates if the relay turns on. Normally, the thermostats are all open, so the relay and alarm bell are off. Under this condition the circuit consumes zero standby current. At 'over-heat' temperatures, on the other hand, one or more of the thermostats closes, and thus turns on the relay and thence the alarm bell. Note that push-button switch S1 is wired in parallel with the thermostats, enabling the circuit to be functionally tested by operating the push-button.

The thermostats used in this and all similar circuits described in this chapter must be n.o. types that close when the temperature exceeds a preset limit. When the thermostats are located in normal living areas they should

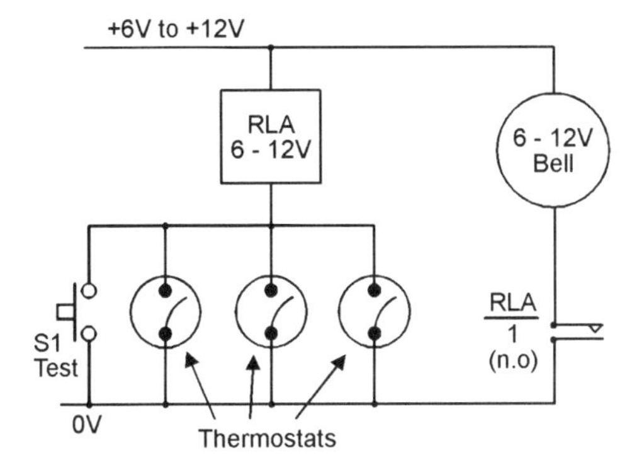

Figure 5.1 *Simple relay-aided non-latching fire alarm*

be set to close at a temperature of roughly 60°C (140°F), but when they are located in unusually warm places – such as furnace rooms or attics – they should be set to close at about 90°C (194°F).

The basic *Figure 5.1* circuit gives a non-latching form of operation. If required, the circuit can be made self-latching by wiring a spare set of n.o. relay contacts in parallel with the thermostats, as shown in *Figure 5.2*. Note that n.c. push-button switch S2 is wired in series with these relay contacts, so that the circuit can be reset or unlatched by momentarily operating S2.

Bell-output fire-alarms can sometimes be activated via an SCR rather than a relay; *Figure 5.3* shows a typical circuit of this type. Here, a normal self-interrupting type of alarm bell is wired in series with the SCR anode, and gate current is provided from the positive supply line via the thermostats and via current-limiting resistor R1. Normally the thermostats are open, so the SCR and bell are off, and the circuit passes only a small leakage current. At

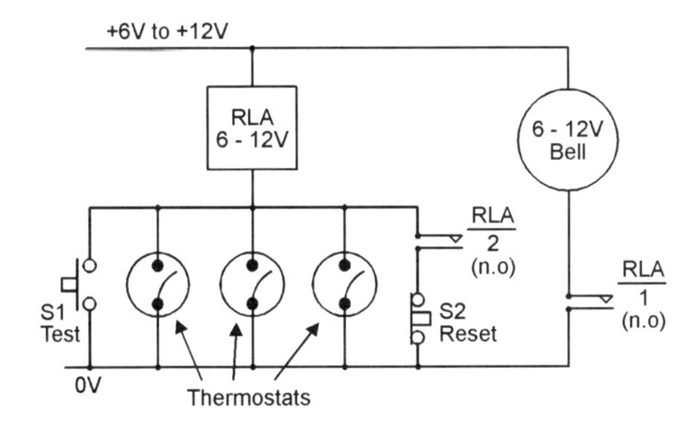

Figure 5.2 *Simple relay-aided self-latching fire alarm*

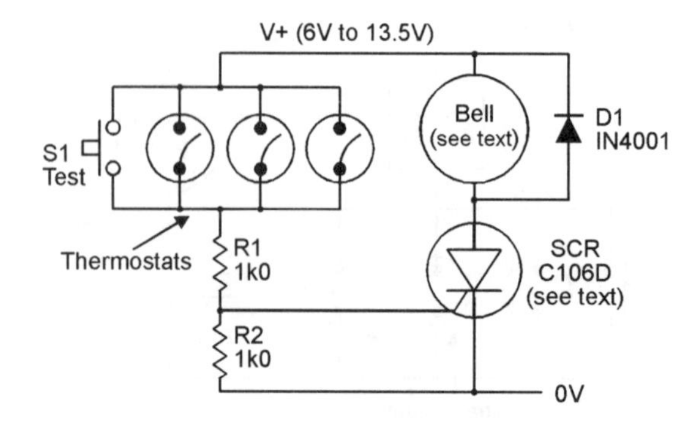

Figure 5.3 *SCR-aided non-latching fire alarm*

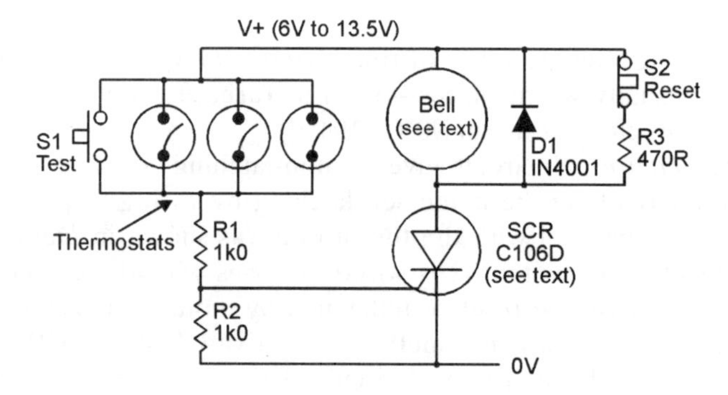

Figure 5.4 *SCR-aided self-latching fire alarm*

high temperatures, however, the thermostats close and apply gate current to the SCR via R1, and the alarm bell turns on.

The basic *Figure 5.3* circuit gives a non-latching form of operation. The circuit can be made self-latching by wiring shunt resistor R3 across the bell as shown in *Figure 5.4*, so that the SCR's anode current does not fall below its latching value when the bell goes into the self-interrupting mode. Note that push-button switch S2 is wired in series with R3 so that the circuit can be reset or unlatched.

It should be noted that the SCR used in the *Figure 5.3* and *5.4* circuits has a current rating of only 2A, so the alarm bell should be selected with this point in mind. Alternatively, SCRs with higher current rating can be used in place of the device shown, but this modification may also necessitate changes in the R1 and R3 values of the circuits. Roughly 1V is dropped across the SCR when it is on, so the circuit's supply line voltage needs to be about 1V greater than the bell's nominal operating voltage.

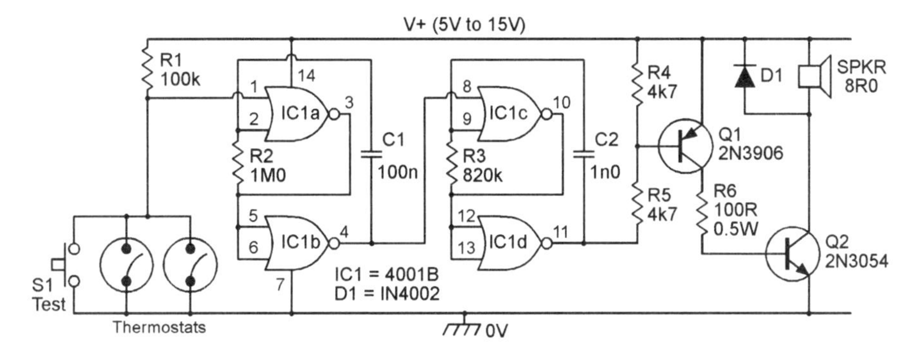

Figure 5.5 *800Hz pulsed-output non-latching fire alarm*

Thermostat fire alarms can be made to generate a siren-type alarm signal directly into a loudspeaker by using the connections shown in *Figures 5.5* or *5.6*. The *Figure 5.5* circuit generates a pulsed-tone non-latching alarm signal, while the *Figure 5.6* circuit generates an 800Hz (monotone) self-latching alarm signal. Both circuits are designed around 4001B CMOS digital IC.

The *Figure 5.5* circuit, which gives an output tone of 800Hz pulsed on and off at a rate of 6Hz, is based on the circuit of *Figure 2.16* combined with a medium-power (see *Figure 2.21*) output stage, and the self-latching *Figure 5.6* circuit is based on the *Figure 2.25* monotone circuit combined with the same medium-power output stage; full descriptions of the operation of these circuits is given in Chapter 2. The *Figure 5.5* and *5.6* circuits should be used with 8R0 horn-type speakers (which are far more efficient than normal speakers), and can use any supply in the range 5V to 15V; they give a mean output power (into the speaker) of about 4.5 watts at 12V or 7.0 watts at 15V.

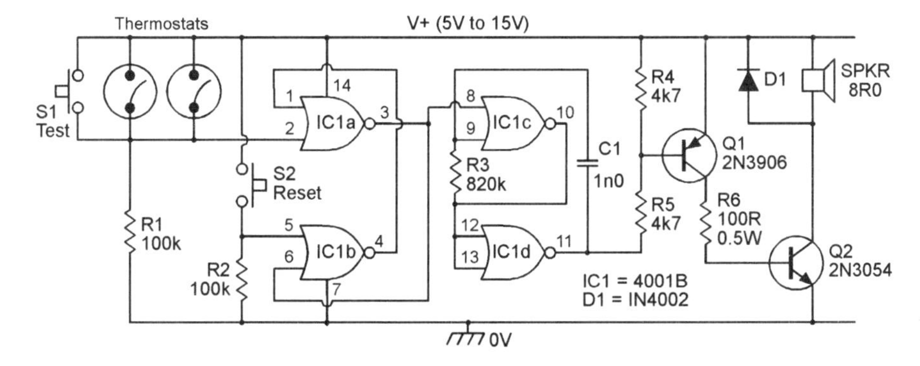

Figure 5.6 *800Hz monotone self-latching fire alarm*

Over-temperature security circuits

Over-temperature security circuits can easily be designed to generate a variety of types of output signal when a monitored temperature rises above a pre-set level, which may range from well below the freezing point of water to well above the boiling point of water. The circuits may be designed to generate an audible signal or to activate some type of relay switch under the 'alarm' condition, and may use one or more thermistors or solid-state devices as their temperature-sensing elements.

Several useful over-temperature security circuits are described in this section. Most of them use inexpensive n.t.c (negative temperature coefficient) thermistors as their temperature-sensing elements. These devices act as temperature-sensitive resistors that present a high resistance at low temperatures and a low resistance at high temperatures. The thermistor circuits described in this and following sections of this chapter have all been designed to work with thermistors that present a resistance of roughly 5k0 at the desired operating temperature; all of these circuits are highly versatile, however, and will in fact work well with any n.t.c. thermistors that present a resistance in the range 1k0 to 20k at the required 'trip' temperature.

Figure 5.7 shows the practical circuit of a simple but very sensitive over-temperature switch that has a relay output. Here, the thermistor (TH1) and RV1–R1–R2 are wired in the form of a simple Wheatstone bridge in which R1–R2 generate a fixed half-supply 'reference' voltage and TH1–RV1 generate a 'variable' output voltage that is inversely proportional to the TH1 temperature and is trimmed (via RV1) so that it almost equals the R1–R2 reference value at the required 'trip' temperature. These two voltages are fed to the input of the type 741 op-amp which is used – in conjunction with transistor Q1 – as the bridge's balance detector and relay driver. The 741 op-amp is used in the open-loop mode in this circuit, and its action is such that

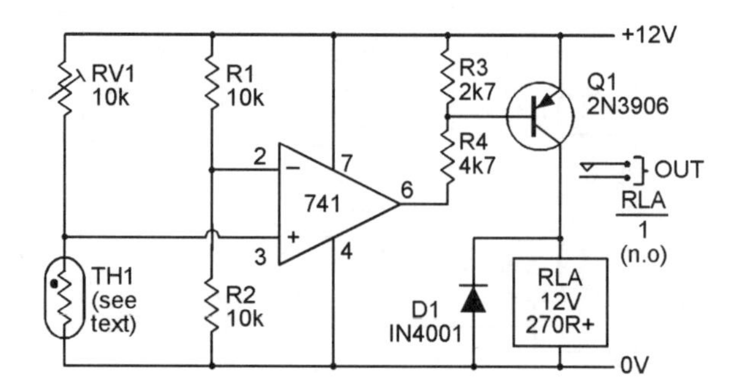

Figure 5.7 *Relay-output precision over-temperature switch*

its pin-6 output is driven low (to negative saturation) if its pin-3 (non-inverting) input is more than a few millivolts negative to the pin-2 (inverting) input, and is driven high (to positive saturation) if pin-3 is significantly positive to pin-2.

Suppose, then, that the bridge is adjusted so that it is close to balance at the desired 'trip' temperature. When the temperature falls below this value the TH1 resistance increases, so the 741's pin-3 voltage rises above that of pin-2, and the pin-6 output voltage thus goes to positive saturation and consequently applies no base drive to Q1; Q1 and the relay are off under this condition. When the temperature rises above the 'trip' value, however, the TH1 resistance decreases and the 741's pin-3 voltage falls below that of pin-2, and the pin-6 output voltage thus goes to negative saturation and applies heavy base drive to Q1; Q1 and the relay are driven on under this condition. Thus, the relay goes on when the temperature rises above the preset level, and turns off when the temperature falls below the pre-set level.

Important points to note about the *Figure 5.7* circuit are that, because it uses the bridge sensing configuration, its accuracy is independent of variations in supply voltage, and that the circuit can respond to TH1 resistance changes of less than 0.1 percent, i.e. to temperature changes of a fraction of a degree. Also note that the basic circuit is quite versatile and can, for example, be converted to a precision under-temperature switch by simply transposing the R1 and TH1 positions, or by transposing the op-amp's pin-2 and pin-3 connections, or by redesigning the Q1 output stage so that it uses an npn transistor in place of the pnp device. Similarly, there are a number of alternative ways of connecting the circuit so that it operates as a precision over-temperature alarm.

One such alternative, shown in *Figure 5.8*, acts as a precision over-temperature alarm with an alarm-bell output. The circuit is similar to that shown in *Figure 5.7*, except that the op-amp's pin-2 and pin-3 connections are transposed, and the op-amp's output is used to drive the gate of an SCR rather than the base of a pnp transistor. The circuit action is such that the

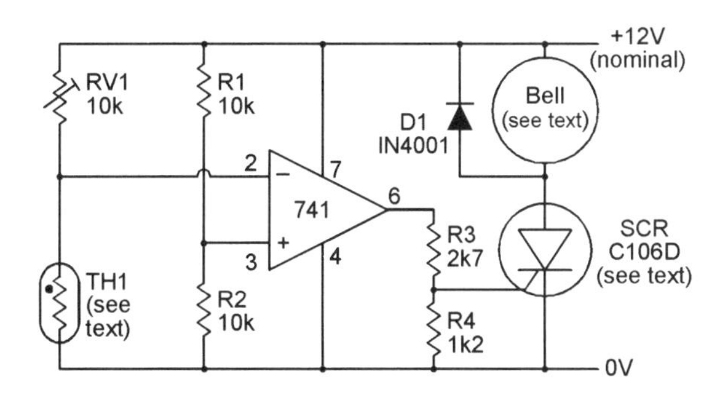

Figure 5.8 *Direct-output precision over-temperature alarm*

op-amp output goes to negative saturation at 'low' temperatures, thus applying zero gate drive to the SCR, but goes to positive saturation at 'high' temperatures, thus driving the SCR and the alarm bell on. The SCR specified in this circuit has a mean current rating of only 2A, so the alarm bell (a self-interrupting type) must be selected with this point in mind.

The *Figure 5.7* and *5.8* circuits use thermistors with nominal trip-level resistance values of 5k0 as their temperature-sensing elements, and these thermistors dissipate several milliwatts of power under actual working conditions. In some special applications this power dissipation may cause enough self-heating of the thermistor to upset its thermal sensing capability. In such cases an alternative type of temperature-sensing device may have to be used.

Ordinary silicon diodes have temperature-dependent forward volt-drop characteristics, and can thus be used as temperature-sensing elements. Typically, a silicon diode gives a forward volt drop of about 600mV at a current of 1mA. If this current is held constant, the volt drop changes by about –2mV for each degree Centigrade increase in diode temperature. All silicon diodes have similar thermal characteristics. Since the power dissipation of the diode is a mere 0.6mW under the above condition, negligible self-heating takes place in the device, which can thus be used as an accurate temperature sensor.

Figure 5.9 shows how general-purpose silicon diode D1 can be used as a thermal sensing element in an op-amp over-temperature relay-switch circuit. Here, zener diode ZD1 is wired in series with R1 so that a constant 5.6V is developed across the two potential dividers formed by R2–RV1 and R3–D1, and a near-constant current thus flows in each of these dividers. A constant reference voltage is thus developed between the R1–RV1 junction and pin-2 of the op-amp, and a temperature-dependent voltage with a coefficient of –2mV/°C is developed between the R1–RV1 junction and pin-3 of the op-amp. Thus, a differential voltage with a coefficient of –2mV/°C appears between pin-2 and pin-3 of the op-amp.

In practice, this circuit is set up by simply raising the temperature of D1 to the required over-temperature trip level, and then slowly adjusting RV1 so that the relay just turns on. Under this condition a differential temperature of about 1mV appears between pin-2 and pin-3 of the op-amp, the pin-3 voltage being below that of pin-2, and Q1 and the relay are driven on. When the temperature falls below the trip level the pin-3 voltage rises above that of pin-2 by about –2mV/°C change in temperature, so Q1 and the relay turn off. The circuit has a typical sensitivity of about 0.5°C, and can be used as an over-temperature switch at temperatures ranging from sub-zero to above the boiling point of water.

Note that the operation of the *Figure 5.9* circuit can be reversed, so that it functions as an under-temperature switch, by simply transposing the pin-2 and pin-3 connections of the op-amp. Also note that dedicated precision temperature-sensing ICs are now available, and are suitable for use in place

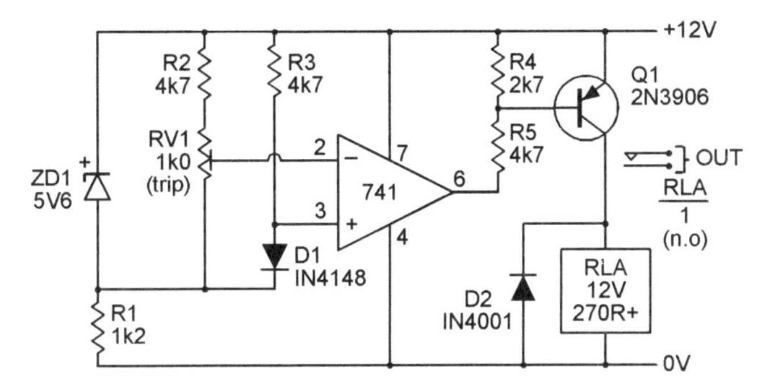

Figure 5.9 *Relay-output over-temperature switch using silicon diode temperature-sensing element*

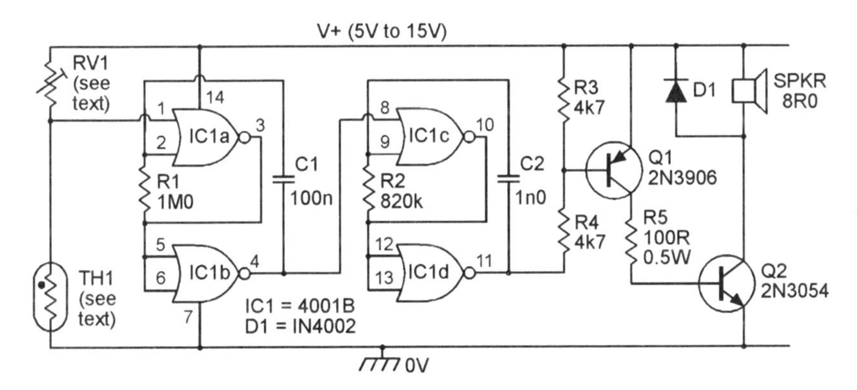

Figure 5.10 *800Hz pulsed-output non-latching over-temperature alarm*

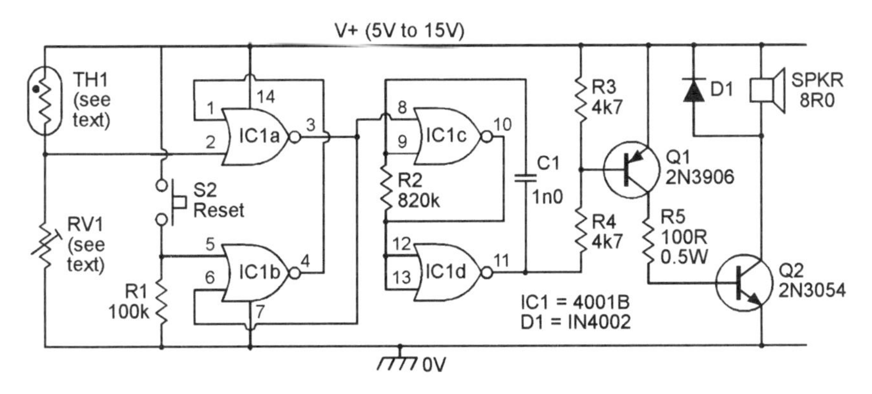

Figure 5.11 *800Hz monotone self-latching over-temperature alarm*

of simple diodes in many high-precision security applications; details of some of these ICs are given later in this chapter.

Finally in this section, *Figures 5.10* and *5.11* show the circuits of a pair of over-temperature alarms that give alarm outputs directly into loudspeakers. The *Figure 5.10* circuit generates a pulsed-tone alarm signal and gives non-latching operation. The *Figure 5.11* circuit generates an 800Hz monotone alarm signal and gives self-latching operation. The *Figure 5.10* and *5.11* circuits are almost identical to the *Figure 5.5* and *5.6* fire-alarm circuits respectively, but have different resistor numbering and have their input activating signals taken from the junction of the RV1–TH1 potential divider rather than from the contacts of the thermostats.

An inherent feature of the *Figure 5.10* and *5.11* circuits is that their 4001B CMOS input stages become enabled or disabled when their input voltages move above or below a certain CMOS 'threshold' value. This threshold voltage is not fixed but is directly proportional to the circuit's supply voltage value. Consequently, these circuits switch from a disabled to an enabled state, or vice versa, when the RV1–TH1 ratios go above or below some particular value. This ratio is independent of the supply voltage, but depends on the threshold value of the individual 4001B IC used in each circuit. The ratio has a nominal value of 50:50, but in reality may vary from 30:70 to 70:30 between individual ICs. In practice, the *Figure 5.10* and *5.11* circuits each turn on when the TH1 temperature exceeds a value that is preset by RV1 (which has a value similar to that of TH1 at the 'trip' level); each circuit has a typical sensitivity of 0.5°C.

Under-temperature security circuits

The over-temperature security circuits of *Figures 5.7* to *5.11* can all be made to give under-temperature operation by making very simple changes to their input connections, as shown in *Figures 5.12* to *5.16*. *Figures 5.12* to *5.14* show

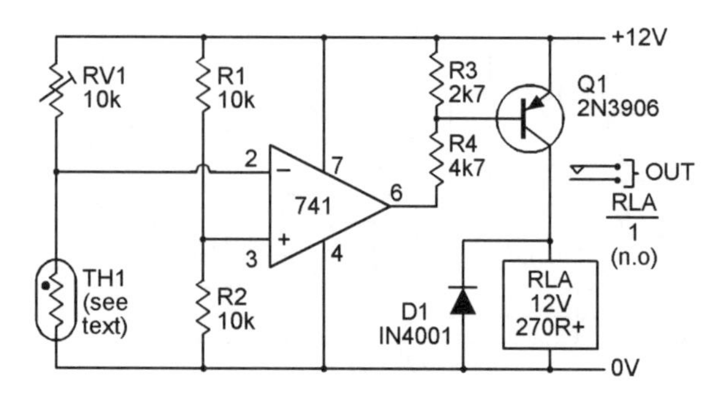

Figure 5.12 *Relay-output precision under-temperature switch*

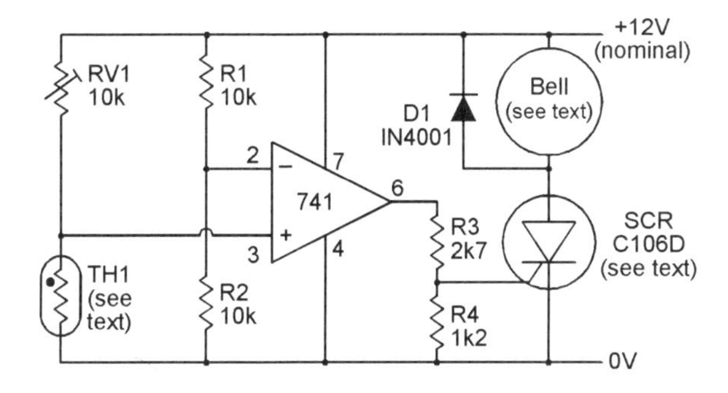

Figure 5.13 *Direct-output precision under-temperature alarm*

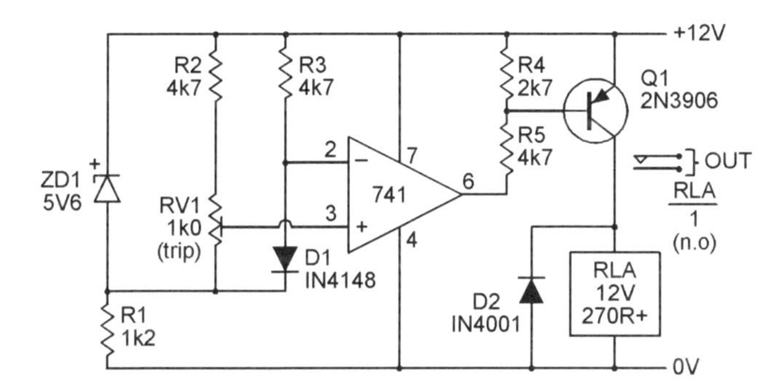

Figure 5.14 *Relay-output under-temperature switch using silicon diode temperature-sensing element*

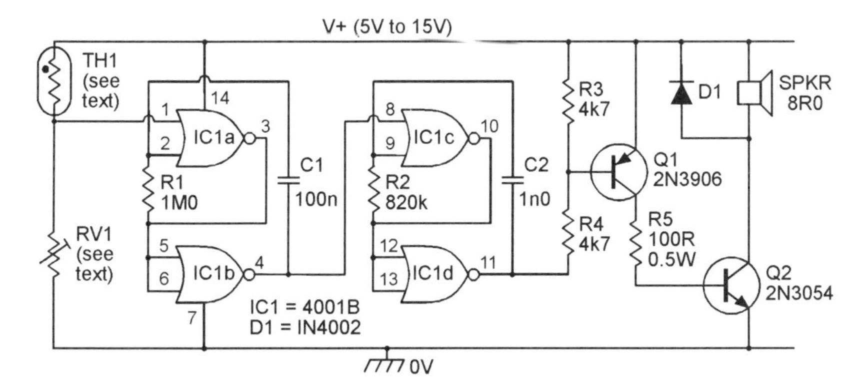

Figure 5.15 *800Hz pulsed-output non-latching under-temperature alarm*

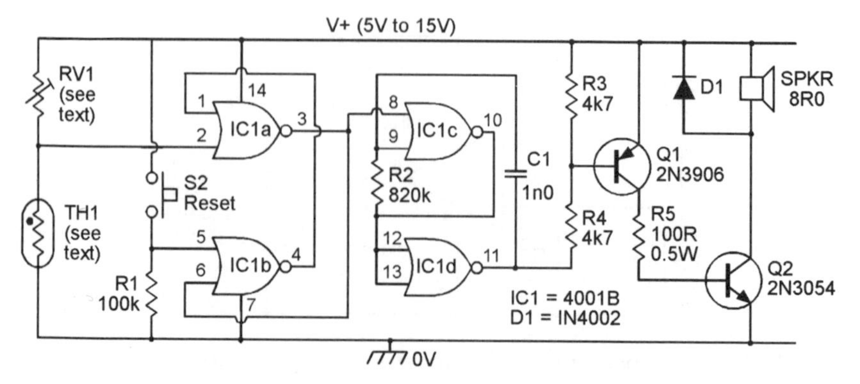

Figure 5.16 *800Hz monotone self-latching under-temperature alarm*

how the circuits of Figures 5.7 to 5.9 can be made to give under-temperature operation by simply transposing the connections to pins 2 and 3 of their op-amps. *Figures 5.15* and *5.16* show how the circuits of *Figures 5.10* and *5.11* can be converted to under-temperature alarms by transposing their TH1 and RV1 positions.

Miscellaneous temperature switches

Each temperature-sensitive circuit shown so far in this chapter activates when a monitored temperature goes either above or below a pre-set level. The present section of the chapter shows three other types of temperature-sensitive security circuit. Two of these circuits activate relays if the temperature deviates from a pre-set level by more that a pre-set amount, and the third activates a relay if two monitored temperatures differ by more than a pre-set amount. In all cases, the relay(s) can be used to operate any type of electrical alarm or slave device.

Figures 5.17 and *5.18* show the circuits of a pair of temperature-deviation switches, which activate if the temperature deviates from a pre-set level by more than a pre-set amount. The *Figure 5.17* circuit has independent over-temperature and under-temperature relay outputs, while the *Figure 5.18* circuit has a single relay output that activates if the temperature goes above or below pre-set levels.

Both of these circuits are made by combining the basic over-temperature and under-temperature circuits of *Figures 5.7* and *5.12*. The right (over-temperature) half of each circuit is based on that of *Figure 5.7*, and the left (under-temperature) half is based on that of *Figure 5.12*. Both halves of the circuit share a common RV1–TH1 temperature-sensing network, but the under-temperature and over-temperature switching levels of the circuits are

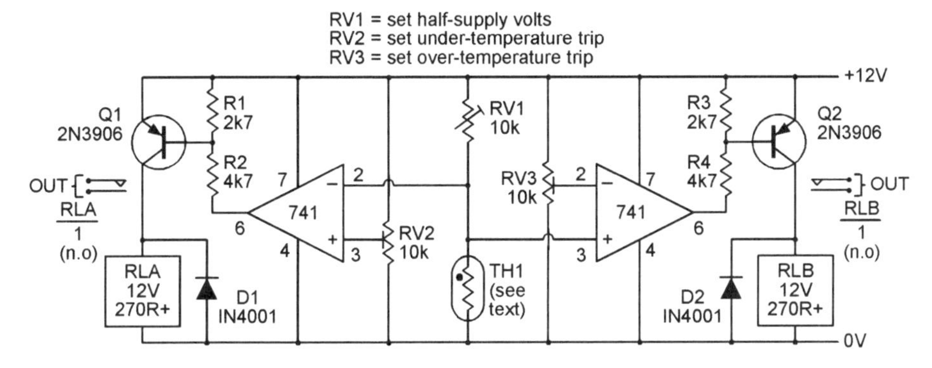

Figure 5.17 *Temperature-deviation switch with independent over/under-temperature relay outputs*

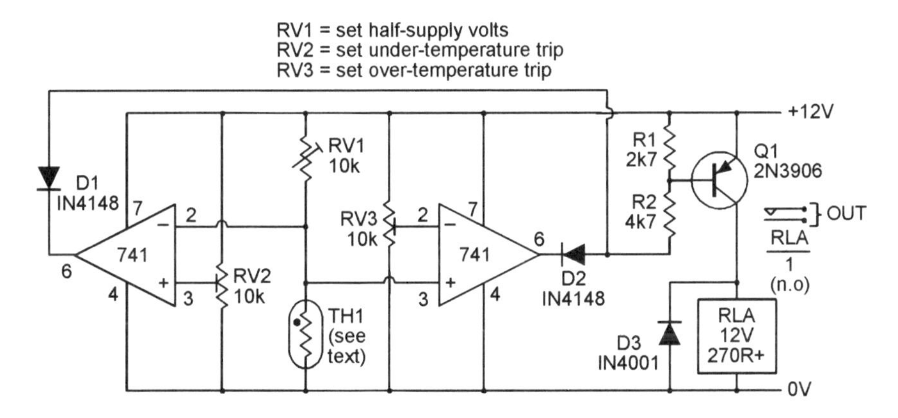

Figure 5.18 *Temperature-deviation switch with single relay output*

independently adjustable. Each of the two op-amp outputs of the *Figure 5.17* circuit are taken to independent transistor–relay output stages, while the two op-amp outputs of the *Figure 5.18* circuit are taken to a single transistor–relay output stage via the D1–D2 gate network. The procedure for setting up the two circuits is as follows.

First, set RV2 and RV3 to roughly mid-travel, then, with TH1 at its normal or mid-band temperature, adjust RV1 so that half-supply volts are developed across TH1. Now fully rotate the RV2 slider towards the positive supply line, rotate the RV3 slider towards the zero volts line, and check that no 'trip' condition is indicated (relays off). Next, reduce the TH1 temperature to the required under-temperature trip value and adjust RV2 so that the appropriate relay goes on to indicate the 'trip' condition. Now increase the TH1 temperature slightly and check that the relay turns off. Finally, raise the TH1

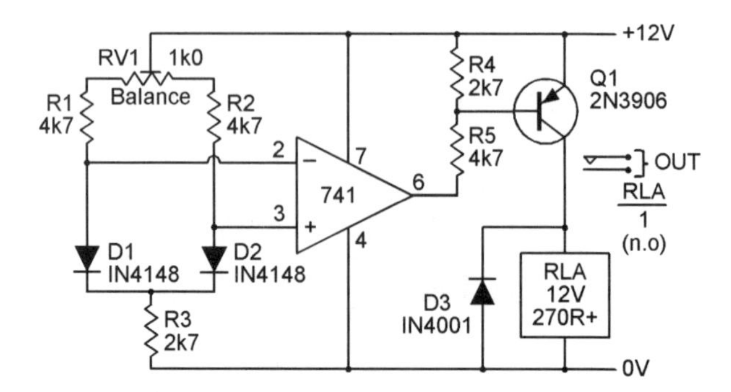

Figure 5.19 *Differential-temperature switch with relay output*

temperature to the required over-temperature trip level and adjust RV3 so that the appropriate relay turns on to indicate the 'trip' condition. All adjustments are then complete, and the circuits are ready for use.

Figure 5.19 shows the circuit of a differential-temperature switch that activates a relay if *two* monitored temperatures differ by more than a pre-set amount. The circuit uses a pair of silicon diodes as temperature-sensing elements, and activates the relay only when the temperature of D1 is more than a pre-set amount greater than that of D2, and is not influenced by the absolute temperatures of the two diodes. Circuit operation is as follows.

General-purpose silicon diodes D1 and D2 are used as temperature-sensing elements. A standing current is passed through D1 from the positive supply rail via RV1–R1 and R3, and a similar current is passed through D2 via RV1–R2 and R3. The relative values of these currents can be adjusted over a limited range via RV1, thus enabling the forward volt drops of the diodes to be equalized, so that they give zero differential output when they are both at the same temperature.

Suppose then that the diode voltages have been equalized in this way, so that zero voltage differential exists between them. If now the temperatures of both diodes are raised by 10°C, the forward voltages of both diodes will fall by 20mV, and zero differential will still exist between them. The circuit is thus not influenced by identical changes in the temperatures of D1 and D2.

Suppose next that the temperature of D2 falls 1°C below that of D1; in this case the D2 voltage will rise 2mV above that of D1, so the op-amp's pin-3 voltage goes positive to that of pin-2, thus driving the op-amp output to positive saturation and holding Q1 and the relay off. Finally, suppose that the temperature of D2 rises 1°C above that of D1; in this case the D2 voltage will fall 2mV below that of D1, so the op-amp output goes into negative saturation and drives Q1 and the relay on. Thus, the relay turns on only when the temperature of D2 is above that of D1. The circuit has a typical sensitivity of 0.5°C.

The above explanation assumes that RV1 is adjusted so that the D1 and D2 voltages are exactly equalized when the two diodes are at the same temperature, so that the relay goes on when the D2 temperature rises a fraction of a degree above that of D1. In practice, RV1 is usually adjusted so that the standing bias voltage of D2 is some millivolts greater than that of D1 at normal temperatures, in which case the relay will not turn on until the temperature of D2 rises some way above that of D1. The magnitude of this differential temperature trip level is fully variable from zero to about 10°C via RV1, so the circuit is quite versatile. The circuit can be set up by raising the temperature of D2 the required amount above that of D1, and then trimming RV1 so that the relay just turns on under this condition.

Precision temperature sensor ICs

Several companies manufacture dedicated temperature-sensing ICs that are suitable for use in place of ordinary silicon diodes in high-precision security applications. Two of the most useful and popular of these ICs are National Semiconductor's commercial-grade LM34CZ and LM35CZ micropower temperature sensors, which are each housed in 3-pin TO-92 packages and

IC number	LM34CZ	LM35CZ
Temp. range	-40°F to +230°F	-40°C to +110°C
Spot accuracy (typ)	±0.8°F at +77°F	±0.4°C at +25°C
Range accuracy (typ)	±1.6°F	±0.8°C
Output scale	10mV/°F	10mV/°C
Supply voltage range	5 - 30V	4 - 30V
Quiescent current	70μA	60μA

Figure 5.20 *Basic parameter values of the LM34CZ and LM35CZ precision temperature-sensor ICs*

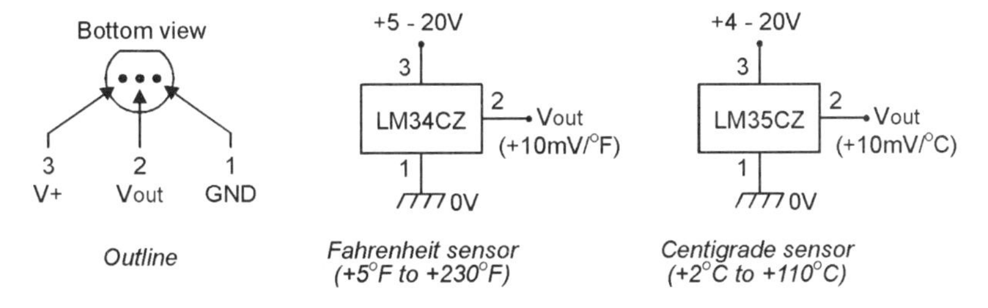

Figure 5.21 *LM34CZ/LM35CZ outline and basic application circuits*

produce output voltages that are linearly equal to +10mV/°F and +10mV/°C respectively. *Figure 5.20* lists the basic specifications of these two ICs, and *Figure 5.21* shows the IC outline and pin notations of the ICs, together with their basic application circuits.

Note in the *Figure 5.21* basic application circuits that – since the circuits use single-ended supplies and the ICs give output voltages that are directly proportional to temperature – these simple designs cannot indicate temperatures that are below zero (in practice, the simple LM34CZ circuit can accurately indicate temperatures down to only +5°F, and the LM35CZ can indicate minimum temperatures of +2°C). If required, the ICs can be made to give full-range outputs (i.e. to give temperature readings down to –40°F or –40°C) by using the dual-supply connections shown in *Figure 5.22*, or by using the single-ended 'simulated dual-supply' connection of *Figure 5.23*.

Also note that – like most micropower ICs – these devices tend to become unstable if their outputs are fed directly to uncompensated capacitive loads greater than a few picofarads. This snag can be overcome by feeding such loads in either of the ways shown in *Figure 5.24*. Finally, note that these ICs can be used to make analogue thermometers by connecting their outputs to

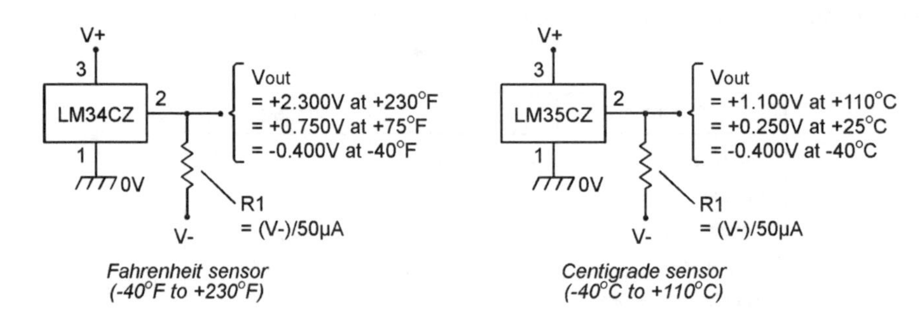

Figure 5.22 *LM34CZ/LM35CZ full-range temperature sensor circuits, using dual power supplies*

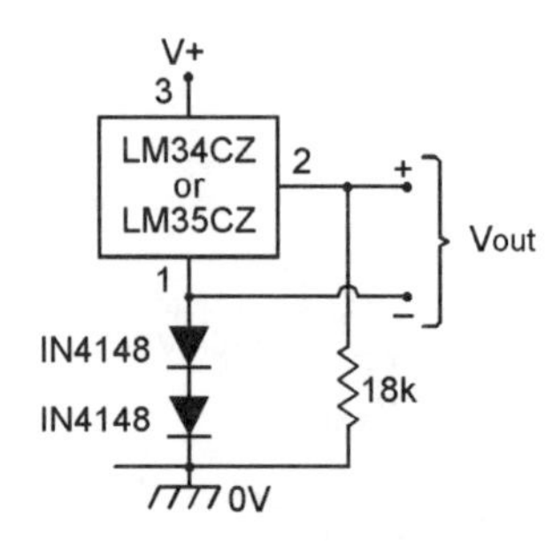

Figure 5.23 *Simulated dual supply full-range temperature sensor, using a single-ended supply*

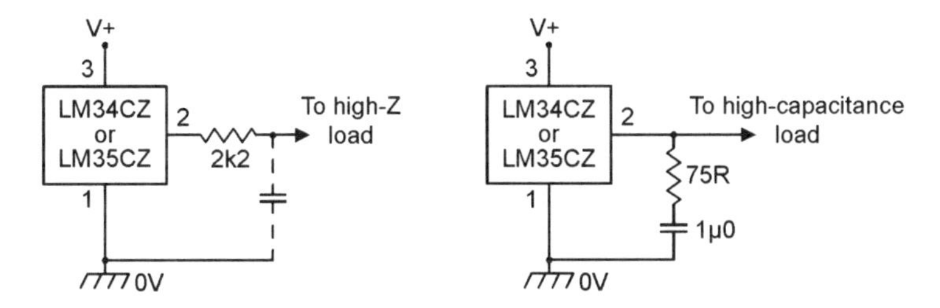

Figure 5.24 *Methods of driving high-capacitance loads*

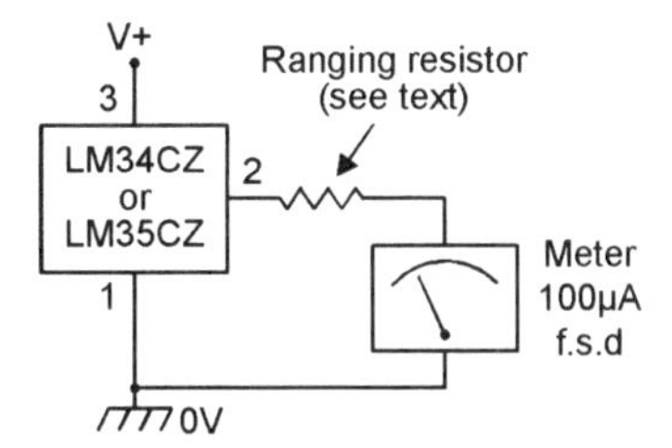

Figure 5.25 *Analogue °F or °C thermometer circuit*

a 100μA moving-coil meter via a suitable 'ranging' resistor, using the connections shown in *Figure 5.25*; this ranging resistor needs an actual value equal to the desired full-scale temperature (in degrees) multiplied by 100R, minus the meter's coil resistance. Thus, to read 120°F full-scale, using a meter with a coil resistance of 3750 ohm, the multiplier needs a resistance of 12k – 3.75k = 8.25k.

6

Instrumentation security circuits

In its simplest form, an instrumentation security circuit is one that activates some kind of electromechanical slave device (such as a relay or solenoid) or alarm unit (such as a siren and/or light strobe) when a monitored voltage, current or resistance value goes above or below some pre-set limit. In practice, this monitored voltage, current or resistance may actually be derived from the output of a transducer that monitors a parameter such as motor speed, liquid flow rate or pressure, or rate of linear movement, etc., and such circuits thus have many practical applications in commerce and in industry. This chapter presents a selection of basic but useful security circuits that can be used to monitor ac or dc voltages or currents, or resistance.

All of this chapter's circuits are designed around CA3140 op-amps (which are usually known simply as '3140' op-amps and can operate from split supply voltage values as low as ±2V, compared to minimum values of ±5V for 741 op-amps) and are designed to operate from single-ended or split 12V supplies, and have outputs that drive the operating coil of a 12V relay, which can activate external circuitry via its contacts. These circuits can, however, easily be modified to drive alternative types of output load.

DC voltage-activated circuits

Figure 6.1 shows the practical circuit of a precision dc over-voltage switch that activates only when the input voltage is greater than some value in excess of at least 5V. Here, the op-amp is used in the open-loop mode as a dc voltage comparator, with a Zener-derived 5V reference applied to the op-amp's non-inverting pin-3 input, and with the test voltage applied between the pin-2 inverting input and ground. The circuit action is such that the op-amp output is positively saturated, and Q1 and the relay are off, when the pin-2 test voltage is fractionally less than the pin-3 5V reference value, and Q1 and the relay go on when the test voltage is greater than the 5V reference value.

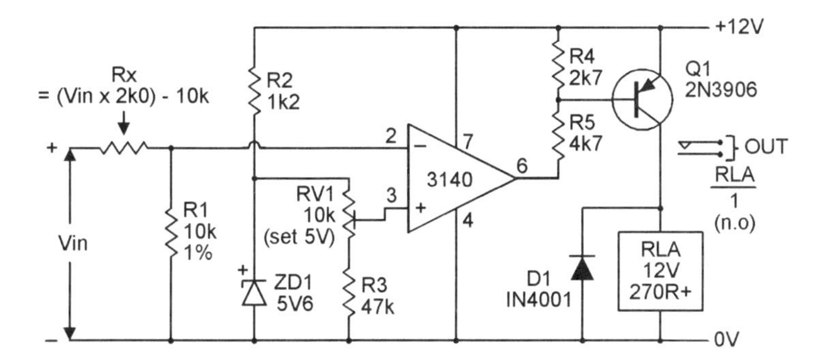

Figure 6.1 *Precision dc over-voltage switch, covering 5V upwards*

Note in *Figure 6.1* that Rx is wired in series between the input test voltage and the 10k (R1) impedance of the inverting input of the op-amp, and enables the circuit to be 'ranged' so that it triggers at any required voltage in excess of the 5V reference value. The Rx value for any required trigger voltage is determined on the basis of (2k0 × V) – 10k. Thus, for 50V triggering, Rx = (50 × 2k0) – 10k = 90k. For 5V triggering, Rx must have a value of zero ohms.

The *Figure 6.1* circuit is very sensitive and exhibits negligible backlash. Triggering accuracies of 0.5 percent can easily be achieved. For maximum accuracy, either the power supply or the Zener reference voltage of the circuit should be fully stabilized.

The *Figure 6.1* circuit can be made to function as a precision under-voltage switch, which turns on when the input voltage falls below a pre-set level, by simple transposing the inverting and non-inverting input pin connections of the op-amp, as shown in *Figure 6.2*. This circuit also shows how the Zener reference supply can be stabilized for high-precision operation. Note in both

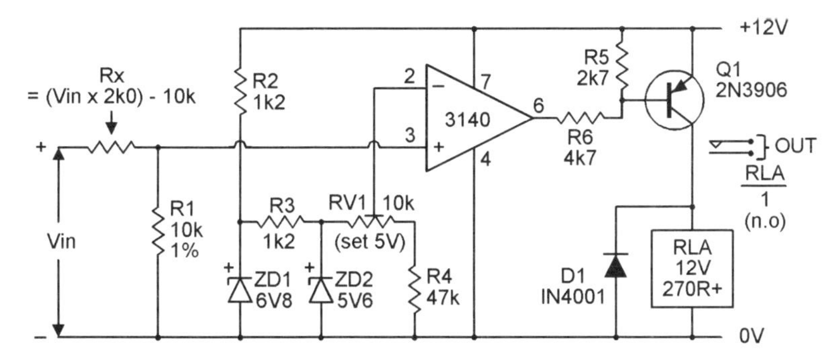

Figure 6.2 *Precision dc under-voltage switch, covering 5V upwards*

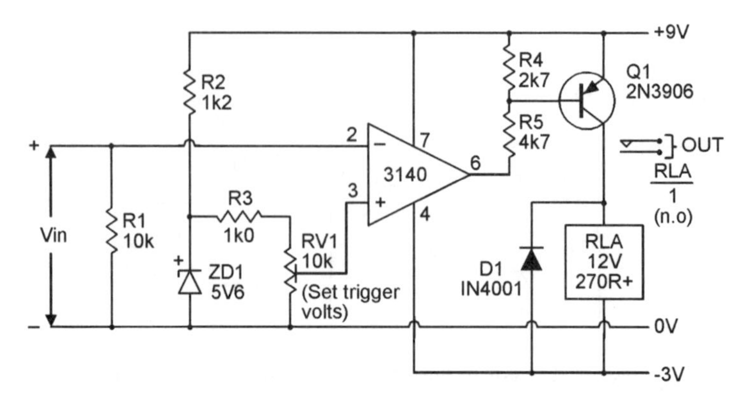

Figure 6.3 *Dual-supply precision dc over-voltage switch, covering 10mV to 5V*

of these circuits that, once 5V has been accurately set via RV1, the final triggering accuracy is determined solely by the accuracies of Rx and R1. In high-precision applications, therefore, these resistors must be stable high-precision types.

Figure 6.3 shows how the *Figure 6.1* circuit can be modified for use as an over-voltage switch covering the range 10mV to 5V. In this case the input voltage is connected directly to the op-amp's inverting input terminal, and a variable reference voltage is applied to its non-inverting input terminal and is adjusted to give the same value as that of the required trigger voltage. The circuit action can be reversed, so that the design acts as an under-voltage switch, by transposing the input pin connections of the op-amp. Note that the *Figure 6.3* circuit uses two sets of supply lines (+9V and –3V), to ensure proper biasing of the op-amp.

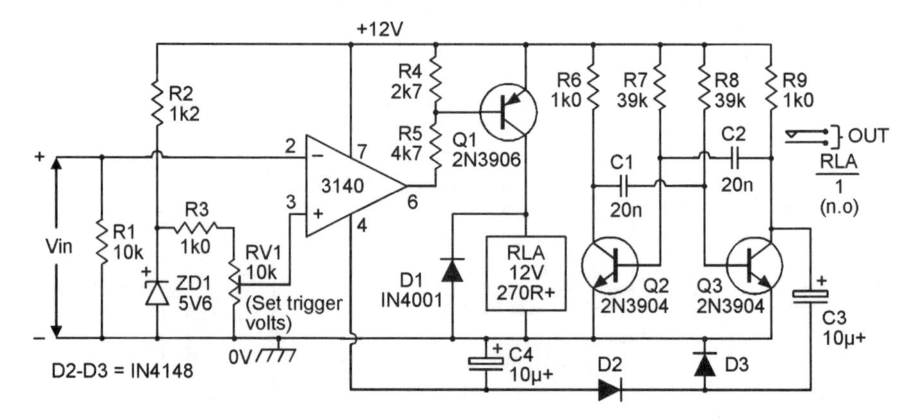

Figure 6.4 *Single-supply precision dc over-voltage switch, covering 10mV to 5V*

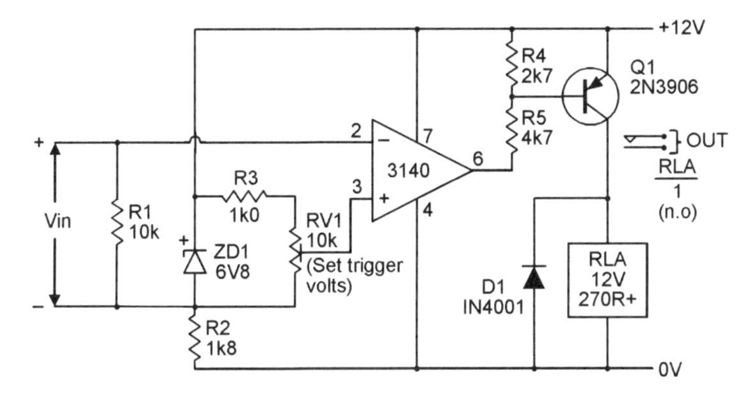

Figure 6.5 *Single-supply dc over-voltage switch, covering 10mV to 5V*

Figure 6.4 shows how the *Figure 6.3* circuit can be adapted for operation from a single set of supply lines. Here, Q2 and Q3 are wired as an astable multivibrator or squarewave generator, and the output of this generator is used to provide a negative supply rail for the op-amp via the voltage-converting and smoothing D2–D3 and C3–C4 network, which actually gives a negative output of about 9V when unloaded, but this output falls to only 3 to 5 volts when it is loaded by connecting it to the pin-4 'negative supply' terminal of the 3140 op-amp.

Finally, *Figure 6.5* shows the circuit of a dc over-voltage switch that covers the range 10mV to 5V and uses a single floating supply. Here, the op-amp is again used as a dc voltage comparator, but its positive supply rail is set at 6.8V via the floating supply and Zener diode ZD1, and its negative rail is set at –5.2V via the ZD1 and R2 combination. The monitored input voltage is fed to the pin-2 inverting input of the op-amp, and the Zener-derived reference voltage is fed to the pin-3 non-inverting input of the op-amp via potential divider R3–RV1. This reference voltage can be varied between roughly 10mV and 5V, and this is therefore the voltage range covered by this over-voltage switch.

AC voltage-activated circuits

The five voltage-activated switches shown in *Figures 6.1* to *6.5* are all designed for dc activation only. All of these circuits can be modified for ac activation by interposing suitable rectifier/smoothing networks or ac/dc converters between their input terminals and the actual ac input signals, so that the ac signals are converted to dc before being applied to the switching circuits.

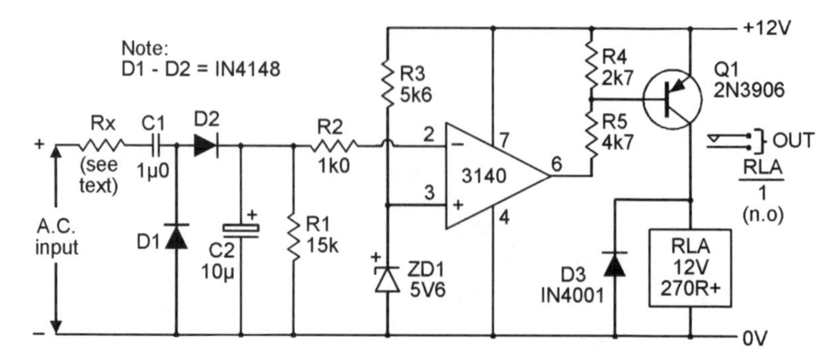

Figure 6.6 *Precision ac over-voltage switch, covering 2.5V upwards*

Figure 6.6 shows the practical circuit of a precision ac over-voltage switch that is designed to work with sinewave signals in excess of 2.5V r.m.s. Here, the ac signal is converted to dc via the voltage-converting and smoothing network formed by Rx–C1–D1–D2–C2–R1, and the resulting dc voltage is applied to the pin-2 inverting input of the op-amp via R2. A Zener-derived 5.6V reference voltage is applied to the op-amp's non-inverting input terminal, and the circuit action is such that – when Rx has a value of zero – Q1 and the relay turn on when the dc voltage on pin-2 exceeds 5.6V.

It is important to note in *Figure 6.6* that the voltage-converting and smoothing network actually gives a dc output voltage that is somewhat less than the *peak* voltage value of a symmetrical sinewave input signal. This type of network cannot be used to measure the r.m.s values of non-symmetrical or pulse-type waveforms. Also note that the action of the *Figure 6.6* circuit can be reversed, so that it acts as an under-voltage switch, by simply transposing the input terminal connections of the op-amp, as shown in *Figure 6.7*.

The circuits of *Figures 6.6* and *6.7* both exhibit a basic input impedance, with Rx reduced to zero ohms, of 15k (= R1), and under this condition a

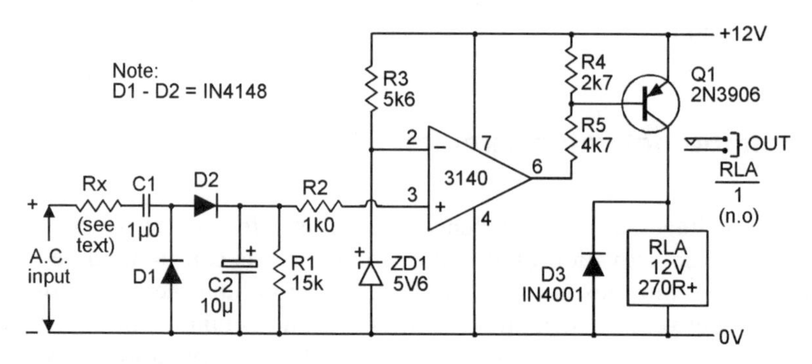

Figure 6.7 *Precision ac under-voltage switch, covering 2.5V upwards*

sinewave of about 2.5V r.m.s. is needed to activate the switch. Consequently, when Rx is given a finite value it acts as a potential divider with the 15k input impedance, and enables the circuits to be triggered at any required ac input level in excess of 2.5V. The Rx value is chosen on the basis of roughly (6k0 × V) – 15k. Thus, if a circuit is to be activated at an input signal level of 10V r.m.s., Rx must have a value of about 45k.

If required, the effective sensitivities of the *Figure 6.6* and *6.7* circuits can be increased, so that they trigger at sinewave input levels substantially less than 2.5V r.m.s., by simply feeding the ac input signals to the inputs of the switching circuits via fixed-gain transistor or op-amp pre-amplifiers.

An alternative way of making an ac-activated over-voltage switch that will trigger at sinewave input voltages well below 2.5V r.m.s. is to feed the voltage to the input of a sensitive dc over-voltage switch (such as that shown in *Figure 6.3*) via an op-amp precision ac/dc voltage converter circuit. Converter's of this type are usually derived from the simple precision half-wave rectifier circuit shown in *Figure 6.8*.

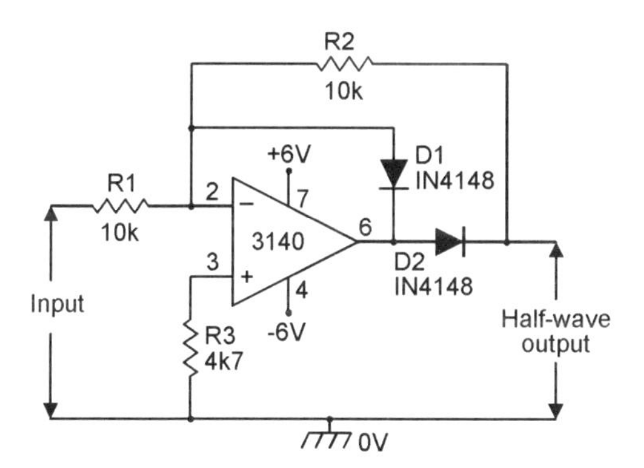

Figure 6.8 *Precision half-wave voltage rectifier circuit*

In *Figure 6.8* the op-amp is wired as an inverting amplifier that has an input impedance equal to the R1 value and has its *maximum* gain set at unity via the R1–R2 ratio, but has its *actual* gain controlled via diodes D1 and D2; the circuit's action is such that it gives zero gain and zero output to positive inputs, but gives unity gain and a positive output to negative input. Thus, when the circuit is fed with a symmetrical sinewave input, it produces a precision positive-going half-wave rectified output voltage; it can be made to give a negative-going half-wave rectified output by simply reversing the two diode polarities.

Figure 6.9 shows how a negative-output version of the above circuit can be combined with an inverting 'adder' to make a precision full-wave rectifier that gives a positive output equal to the absolute value of the input signal.

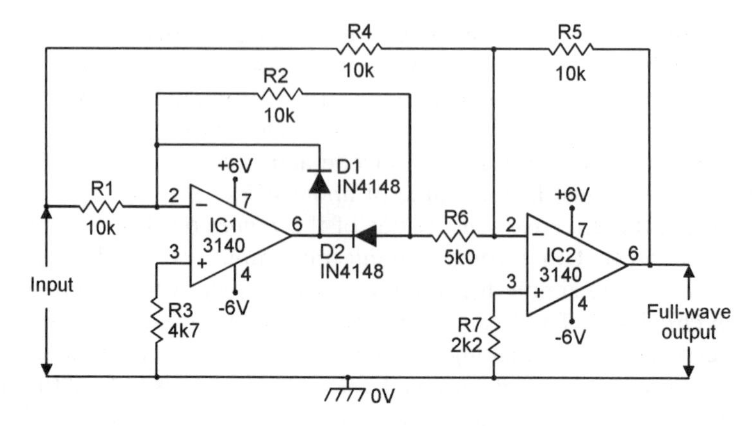

Figure 6.9 *Precision full-wave voltage rectifier circuit*

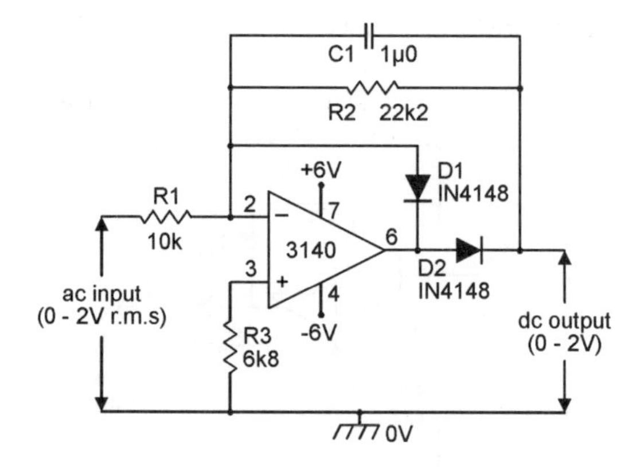

Figure 6.10 *Precision half-wave ac/dc voltage converter*

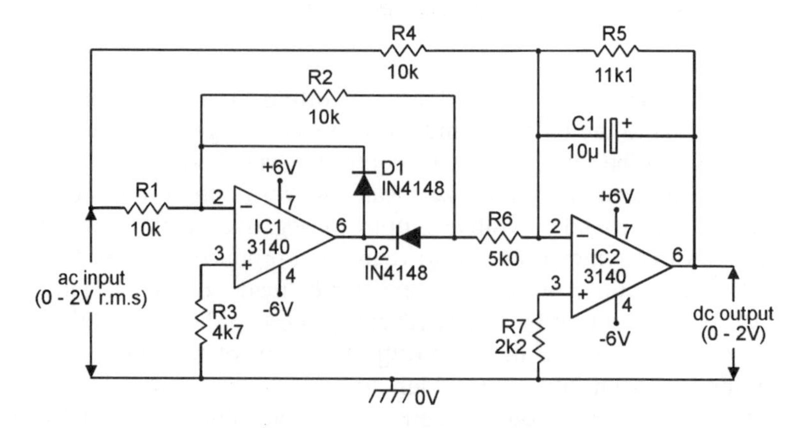

Figure 6.11 *Precision full-wave ac/dc voltage converter*

The *Figure 6.8* and *6.9* circuits can be made to function as precision ac/dc converters by first providing them with voltage-gain values suitable for form-factor correction, and by then integrating their outputs to give the ac/dc conversion, as shown in *Figures 6.10* and *6.11* respectively.

In the precision half-wave ac/dc converter of *Figure 6.10* the circuit is given a voltage gain of ×2.22 via R2/R1, to give sinewave form-factor correction, and integration is accomplished via C1–R2. Note that this circuit has a high output impedance, and the output must be buffered if fed to low-impedance loads.

In the precision full-wave ac/dc converter of *Figure 6.11*, the circuit has a voltage gain of ×1.11 to give form-factor correction, and integration is accomplished via C1–R5. This circuit has a low-impedance output.

Current-activated circuits

Each of the five dc voltage switch circuits of *Figures 6.1* to *6.5* can be used as a DC (power-level) current switch by simply feeding the monitored DC current to the input of the voltage switch via a current-to-voltage converter. A suitable converter circuit is shown in *Figure 6.12*.

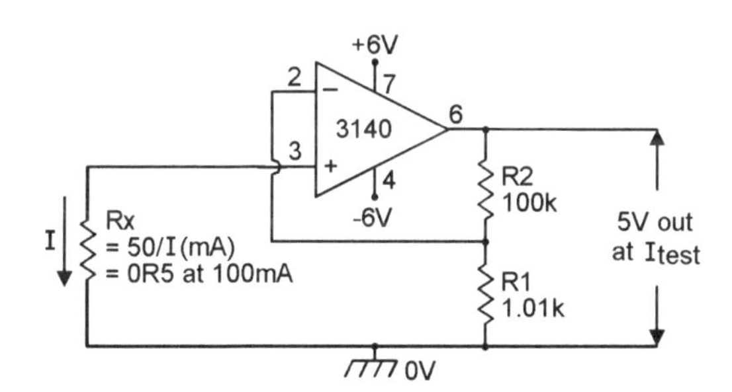

Figure 6.12 *DC current-to-voltage converter*

In *Figure 6.12*, the op-amp is wired as a non-inverting ×100 voltage amplifier, with its gain controlled by the ratios of R1 and R2. The test current is passed through input resistor Rx, which has its value chosen so that 50mV is developed across it at the required trigger current, thus giving 5V output from the op-amp under this condition; this 5V output is used to trigger the external voltage switch. The circuit's Rx value is selected on the basis of

Rx = 50/I(mA)

where I(mA) is the desired trigger current in milliamps. Thus, Rx needs a value of 0R5 at a trigger current of 100mA, or 0.05 ohms at a trigger level of 1A, and so on.

Note that the *Figure 6.12* type of converter circuit can also function as an AC current-to-voltage converter, and can be used to make an AC current switch by connecting its output to the input of a suitable ac voltage switch circuit.

If required, an op-amp circuit can be built specifically for use as a DC over-current switch by using the circuit shown in *Figure 6.13*. Here, Rx is again used to develop 50mV at the desired test current level, and this voltage is applied to the inverting pin of the op-amp. A Zener-derived reference of approximately 50mV is applied to the non-inverting input terminal of the op-amp; this reference voltage can be adjusted over a limited range via RV1, thus providing a limited control of the circuit's sensitivity.

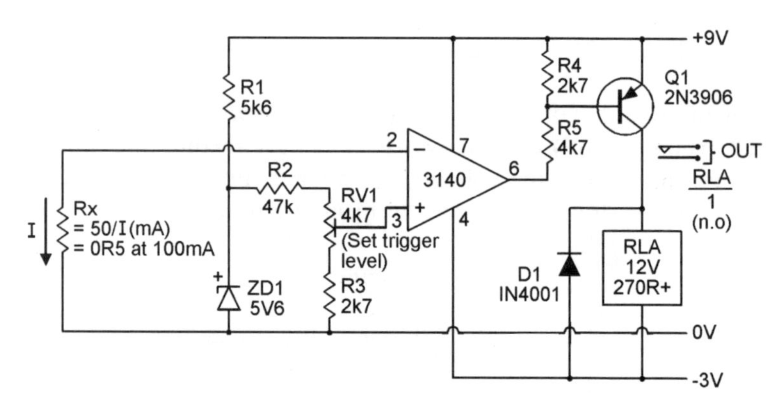

Figure 6.13 *DC over-current switch*

Thus, the *Figure 6.13* switch circuit's relay turns on when the current-derived input voltage exceeds the 50mV reference voltage. The circuit's action can be reversed, so that it acts as an under-current switch, by simply transposing the op-amp's two input terminal connections. In either case, the value of monitor resistor Rx is chosen on the basis of

$$Rx = 50/I(mA)$$

where I(mA) is the desired trigger current in milliamps. Thus, Rx needs a value of 0R5 at a trigger current of 100mA, or 0.05 ohms at a trigger level of 1A, and so on.

Resistance-activated circuits

Figure 6.14 shows the practical circuit of a precision under-resistance switch that turns on when the value of a monitored resistance falls below a specific

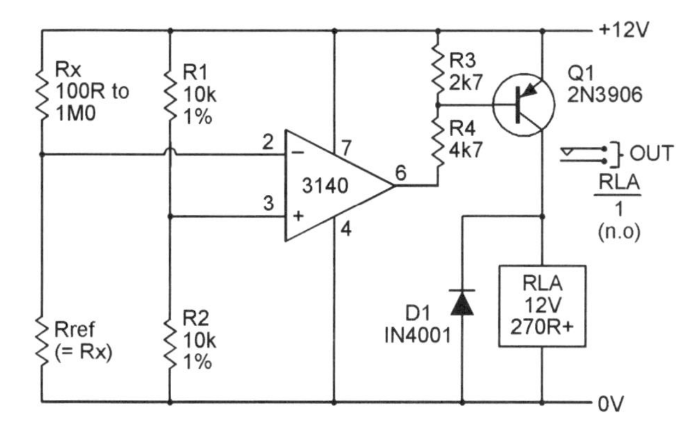

Figure 6.14 *Precision under-resistance switch*

value. Here, the op-amp is again used as a voltage comparator, with its output fed to the coil of relay RLA via transistor Q1, but in this case the voltage on the op-amp's non-inverting input pin is set at half-supply volts via potential divider R1–R2, and the voltage on the inverting input pin is determined by the ratios of Rx and Rref. In effect, these four resistors are wired as a Wheatstone bridge, and the circuit action is such that the relay turns on when the value of Rx falls below that of reference resistor Rref, i.e. when the bridge goes out of balance in such a way that the voltage on the op-amp's inverting input terminal rises above that on its non-inverting input terminal.

In this circuit, Rref must have the same value of the desired Rx 'trigger' resistance value, which can have any value in the range 100R to 1M0; the minimum usable resistance value is dictated by the current-driving capability of the circuit's power supply (and the power dissipation limits of Rref and Rx), and the maximum value is restricted by the shunting effect that the op-amp's input impedance and leakage impedance has on the effective value of Rref.

The accuracy of the above circuit is (within sensible limits) quite independent of variations in power-supply voltage, and the 'switch' circuit is capable of responding to changes of less than 0.1 percent in the value of Rx. The *actual* accuracy of the circuit is determined by the precisions of R1–R2 and Rref, and in worst-case terms is equal to the sum of the tolerances of these three resistors, i.e., it equals ±3% if ±1% resistors are used.

The action of the above circuit can be reversed, so that it acts as a precision over-resistance switch, by simply transposing the input pin connections of the op-amp, as shown in *Figure 6.15*. This circuit also shows how the basic accuracies of both designs can be improved by adding an RV1 'set balance' control to the R1–R2 potential divider chain. This control enables the bridge to be very precisely balanced (even when using ±5% R1–R2 components) so that the circuit 'switches' when the value of Rx varies from the marked value

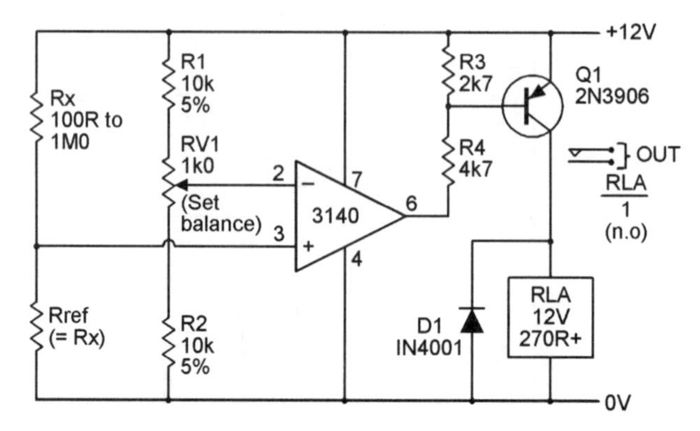

Figure 6.15 *Precision over-resistance switch*

of Rref by only ±0.1% or so. In this case, the true accuracy of the circuit is equal to the tolerance of Rref ±0.1%.

A led 'flasher' output circuit

All of the instrumentation security circuits shown so far in this chapter are designed to activate a relay, via Q1, under the 'switch on' condition. In practice, any of these circuits can be made to switch an alternative type of output load, such as a LED indicator and/or a buzzer or siren, etc., by simply connecting such a load to the output of the circuit's op-amp in such a way that the load is activated when the op-amp's output is switched low (rather than high). *Figure 6.16*, for example, shows an alternative output circuit that can be used with any of the *Figure 6.1* to *6.15* designs to activate an atten-

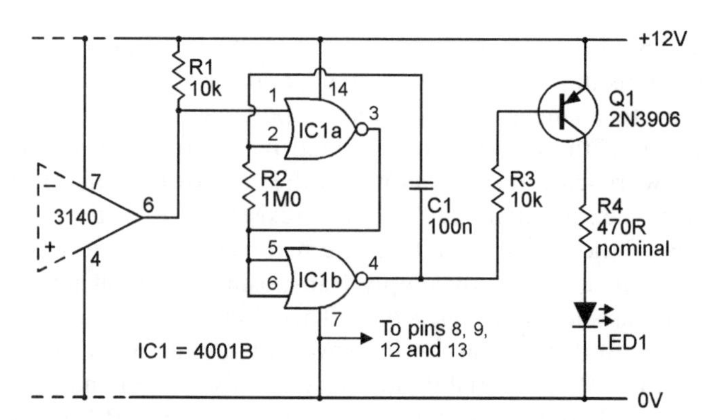

Figure 6.16 *An optional 6Hz LED flasher output circuit*

tion-grabbing LED-flasher circuit under a monitored 'fault' condition.

In *Figure 6.16*, two of a 4001B CMOS IC's gates are wired as a gated 6Hz astable multivibrator that is gated by the op-amp's output and is used to switch a LED on and off via R3 and Q1. Normally, when the IC's pin-1 terminal is held high by the op-amp's output, the astable is disabled and the LED is off, but when pin-1 is pulled low the astable is gated on and under this condition pulses the LED on and off at an approximate 6Hz rate (if desired, this pulse rate can be altered by changing the R2 value).

This LED flasher circuit uses the existing power supplies of the instrumentation security circuit, and is used in place of the normal circuit's Q1 (etc.) relay-driving network, with pin-14 going to the +ve rail and pin-7 to the low-voltage (0V or –ve) supply rail; pin-1 of the 4001B connects directly to the pin-6 output terminal of the circuit's op-amp. The LED's 'on' current value is controlled by the R4 value.

7

Automobile security circuits

Modern automobiles virtually bristle with various 'security' devices and gadgets, which are designed to enhance the vehicle's safety, reliability, mechanical efficiency, and immunity to theft. If you own a reasonably modern (post 1994) automobile, it is probably already so well equipped with good security devices that you will not need to add any more. If, on the other hand, your vehicle dates back well beyond 1989, you may be able to gain by fitting one or more of the simple security circuits that are described in this chapter, or by using one or more of the various commercial 'add-on' vehicle security units that are widely available.

To help illustrate some of the advances that have taken place in automobile security in recent years, let me cite two of my own personal experiences. At the time of writing (1998) I own a perfectly standard 1995, bottom-of-the-range, European-built 4-door family saloon, and before that I owned a 1986 version of the same basic vehicle. The 1986 vehicle was (as supplied from the factory) fitted with no anti-theft devices other than simple key-controlled door, steering and ignition locks, had a traditional petrol engine that used a normal carburettor and had a manual choke, and had 2-speed windshield wipers but was otherwise completely devoid of modern 'security/safety' devices.

By contrast, my 1995 vehicle is fitted – as standard equipment – with a central-locking system that controls the security of its four doors and can be activated manually or via an infra-red remote-control key-fob, is fitted with three different anti-theft systems, has a smart electronically-controlled fuel-injection petrol engine with an automatic choke, has multi-speed windshield wipers, has a timed auto-turn-off rear-screen heater, and has a 'lights-are-on' reminder that automatically activates a bleeper if I try to leave the vehicle while its lights are still turned on.

My 1995 vehicle's three anti-theft systems are additional to its normal key-controlled door, steering and ignition locks, and consist of a basic switch-activated alarm unit that activates if any of the vehicle's doors or its trunk (boot) or bonnet (hood) are illegally opened, plus a dual narrow-beam short-range ultrasonic unit that triggers an alarm if anyone pushes an arm through either of the two front-side windows while the alarm is turned on, plus a key-pad engine immobilizer that only enables the electronically-controlled fuel-injection system if the correct ignition key is used and a personal 4-digit code is punched into the key-pad before trying to start the engine.

This chapter presents a variety of automobile add-on security circuits that are particularly suitable for use in most older vehicles that are fitted with 12V electrical systems. Its main section starts by describing vehicle anti-theft principles, then presents a few simple but useful anti-theft circuits, and then goes on to describe various simple circuits that can be used to enhance a vehicles safety or reliability.

Vehicle anti-theft principles

Vehicle thieves

Vehicle thieves come in three basic types, and can be categorized as either car burglars, drive-away thieves, or hoist-away thieves. Car burglars are mainly interested in stealing saleable articles from vehicles; usually, they enter the vehicle by smashing a side window and then opening the side door and grabbing any valuables that are lying around; they work fast, do expensive damage to the vehicle, and are not usually deterred by burglar alarms. Some car burglars specialize in stealing vehicle parts such as wheels (and their easily-sold tyres) and radios. It is almost impossible to protect modern vehicles against really determined car burglars.

Drive-away thieves simply break into the vehicle and drive it away for temporary pleasure or for use in some criminal enterprise or for resale in either complete or broken-down form. Most drive-away thieves can be deterred by a well-designed burglar alarm and a good and *sensible-fitted* immobilizer unit.

Hoist-away thieves steal a complete vehicle by either towing it away or hoisting it onto the back of a truck; the vehicle is then (usually) deposited in an enclosed workshop, where it is processed ready for resale through a local, national, or international dealing organization. Most hoist-away thieves only steal vehicles with a good resale value, and are rarely detected by burglar alarms or inconvenienced by immobilizers. Vehicles can be given fairly good protection against this type of theft only by fitting the vehicle with a modern (and expensive) wireless ‘tracker’ system, as described in the next section of this chapter.

Anti-theft devices

Vehicle anti-theft devices come in four basic types. The first and simplest of these are mechanical ‘theft’ deterrents such as steering locks and lockable filler caps and wheel nuts, etc. Note that most built-in steering locks (of the type that are operated via the vehicles ignition key) are fairly weak, and can often be broken by an experienced car thief – using brute force – in just a few seconds. Add-on steering locks are usually very strong and are excellent deterrents against drive-away car thieves, but are cumbersome to use.

The second type of anti-theft device is the 'immobilizer', which is intended purely to reduce a thief's chances of starting or driving away a target vehicle; it offers no protection against the car burglar or against haul-away thieves. Immobilizers usually consist of some type of electronic or electromechanical 'switch' that is wired into some part of the electrical wiring of the vehicles engine, so that the engine will only operate if the 'switch' is disabled.

The third type of anti-theft device is the true burglar alarm, which sounds an alarm (and perhaps also immobilizes the vehicle's engine) if any unauthorized person tampers with the vehicle or tries to force their way into it. These alarms may be activated in any of several basic ways. The most popular way is via microswitches that operate when any of the car doors, hood or trunk are opened. Microswitch-activated alarms are fairly inexpensive, very reliable, and give excellent anti-theft protection.

Another way of detecting an intrusion is via the small volt drop that takes place across the vehicle's battery when a door-, hood- or trunk-activated courtesy light turns on, or when the ignition is turned on. These so-called 'voltage sensing' alarms give the same degree of anti-theft protection as the microswitch types of alarm system, but are generally more expensive and less reliable.

Some vehicle alarms are activated by a vibration or movement detector, which reacts when a vehicle is entered or moved. Most of these systems have poor reliability, and can be false-triggered by natural movements caused by vehicles passing close by, or by strong gusts of wind, or by potential thieves deliberately rocking the vehicle in an effort to destroy the owner's trust in the system.

Some vehicle alarms are ultrasonic 'doppler shift' types, in which an ultrasonic transmitter (Tx) and a matching receiver (Rx) are aimed into the vehicle's driver/passenger compartment. If anyone enters any part of the compartment they cause a doppler shift in the Rx signal, and this activates the alarm. In practice, similar doppler shifts can be caused by draughts blowing through poorly closed windows or ventilators, and many such units thus have very poor reliability. In most modern vehicles, ultrasonic units are only used to give short-range protection of small areas, such as front-side windows, and are used in conjunction with conventional microswitch-types of alarm unit, rather than as stand-alone alarms.

Practical vehicle burglar alarm systems may be designed to be switched on and off either from within the vehicle, or from outside. Up until the late 1980s, most vehicle burglar alarm systems were switched from within the vehicle. Internally switched burglar alarms must incorporate exit and entry delays of about 20 seconds, to enable the owner to leave or enter the vehicle without sounding the alarm. This entry delay makes the vehicle very vulnerable to car burglars, who can easily carry out a snatch-theft in less than 20 seconds and thus escape before the vehicle's alarm sounds.

By contrast, externally switched vehicle burglar alarms sound off the instant that any car door starts to open, and thus give better protection against car burglars. Most modern cars are fitted with a central-locking system that protects all doors, and this system and the burglar alarm are controlled (turned on and off) externally via the driver's door key or via a wireless or IR remote-control key-fob. If you want to give this kind of protection to an older vehicle you can do so by buying a fit-it-yourself central locking system and remote-controlled burglar alarm, etc., from a specialist dealer.

A cheaper way of fitting an externally switched burglar alarm to an older vehicle is to simply fit microswitches to all car doors and to the hood and trunk, to use these to activate one of the simple anti-theft alarms described later in this chapter, and to turn the alarm on and off via a hidden toggle-switch or a prominent key-switch that is fitted to the outside of the vehicle.

The fourth and final type of vehicle anti-theft device is the wireless 'tracker' system, which is meant to be used in conjunction with a normal burglar alarm/immobilizer system and is used solely to quickly (often in less than half an hour) help track down the vehicle if it is stolen, even if it is hidden away in an enclosed workshop. A 'tracker' is a special wireless receiver/transmitter unit that normally lies dormant, hidden away in the vehicle. If the vehicle's owner reports the vehicle's theft to the police (or some other law-enforcement authority), they will wireless-transmit that vehicle's special 'tracker' code number which – if received – will activate the tracker's transmitter, thus enabling the vehicle to be quickly found with the aid of suitable wireless-location equipment.

Device type	Car burglar	Drive-away thief	Hoist-away thief
Immobiliser	Nil	Good	Nil
Internally-switched microswitch-activated alarm	Nil	Good	Nil
Externally-switched microswitch-activated alarm	Fair	Good	Nil
Externally-switched voltage-sensing alarm	Fair	Good	Nil
Externally-switched ultrasonic alarm	Fair	Fair	Nil
Externally-switched vibration alarm	Fair	Fair	Fair
'Tracker' unit	Nil	Nil	Good

Figure 7.1 *Comparative table showing levels of protection given by various anti-theft devices against three types of thief*

The comparative table of *Figure 7.1* shows the degree of protection that different types of vehicle anti-theft device offer against different types of thief. Thus, immobilizers give good protection against drive-away thieves but give no protection against car burglars or hoist-away thieves, while externally switched microswitch-activated and voltage-sensing alarms give good protection against all except hoist-away thieves.

Practical immobilizer circuits

An immobilizer can greatly reduce a thief's chances of starting or driving away a target vehicle. Simple immobilizers can take the form of a secret switch (or a pair of remotely-controlled relay contacts) wired into some vital electrical part of the vehicle's engine; *Figures 7.2* to *7.5* show some basic circuits of this type.

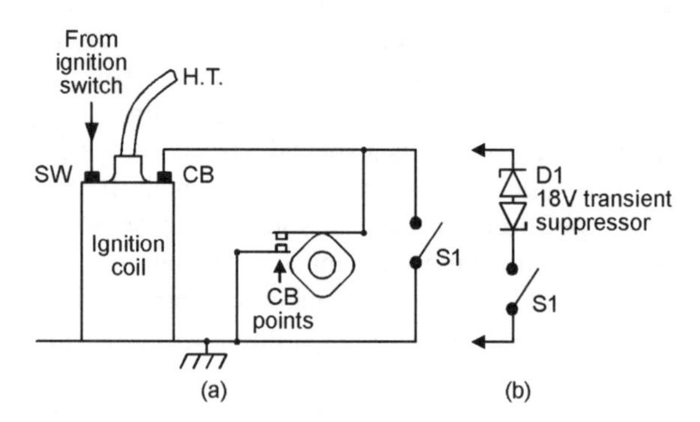

Figure 7.2 *Contact-breaker immobilizer, operates when switch is closed; (a) is basic circuit, (b) is improved circuit*

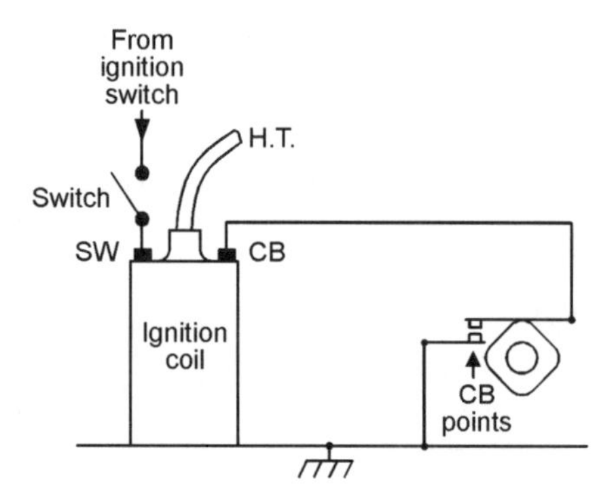

Figure 7.3 *Ignition immobilizer, operates when switch is open*

Figures 7.2 and *7.3* show how immobilizers can be wired into the engine's ignition system. In *Figure 7.2(a)*, switch S1 is wired across the engine's contact-breaker (CB) points; when S1 is open the ignition operates normally, but when it is closed the CB points are shorted out and the engine is unable to operate. This simple circuit gives excellent protection, particularly if the wiring is carefully concealed at the CB end, but S1 must be able to handle the coil's high operating current when it is closed, and be able to withstand the CB point's typical 600V peak-to-peak 'ringing' voltages when the engine is operating normally.

The improved immobilizer circuit of *Figure 7.2(b)* does not suffer from the above snags. Here, an inexpensive and easily available 18V transient suppressor diode (D1) – which acts like a pair of back-to-back Zeners – is wired in series with S1; this device passes no current until its applied voltage exceeds 18V, and acts like a near-short to voltages greater than 18V. Thus, in this circuit, S1 passes zero DC current when it is closed, but kills the ignition systems vital 'ringing' voltages when an attempt is made to start the engine. This circuit gives superb anti-theft protection.

In the *Figure 7.3* circuit the immobilizer switch is wired in series with the vehicle's ignition switch, so that the engine operates only when the switch is closed. The protection offered by this circuit is not quite as good as that of *Figure 7.2*, since a moderately skilled thief can easily by-pass the immobilizer and ignition switches by simply hooking a wire from the battery to the SW terminal of the ignition coil.

Figure 7.4 shows how a heavy-duty immobilizer switch can be wired into the vehicle's electric starter system, so that the starter only operates if this switch is closed. This system gives better protection than that of *Figure 7.3*, but is not as good as *Figure 7.2* because the starter solenoid can be operated manually on many old vehicles, and also because the starter and immobilizer switches can be by-passed by a single length of wire.

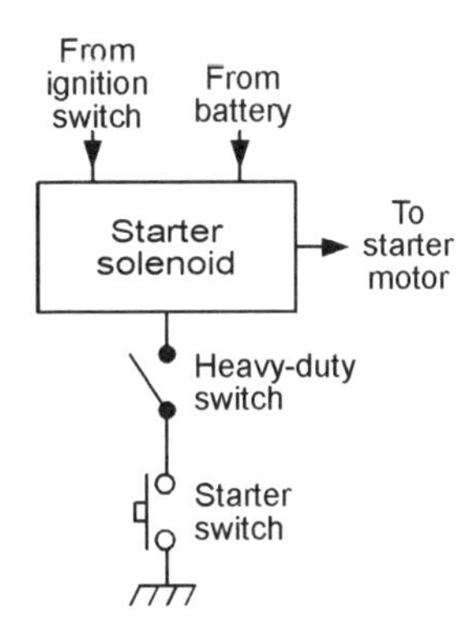

Figure 7.4 *Starter-motor immobilizer, operates when switch is open*

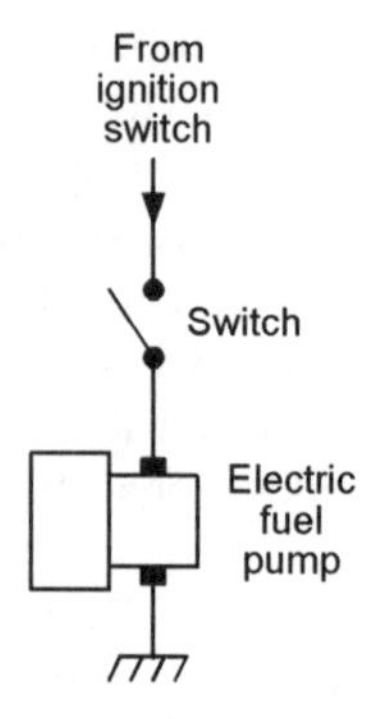

Figure 7.5 *Fuel-pump immobilizer, operates when switch is open*

Finally, *Figure 7.5* shows how an immobilizer switch can be wired in series with the electric fuel pump on suitable vehicles, so that the pump only operates when this switch is closed. Note that this system lets a thief start the engine and drive for a short distance on the carburettor's residual fuel, which is quickly used up.

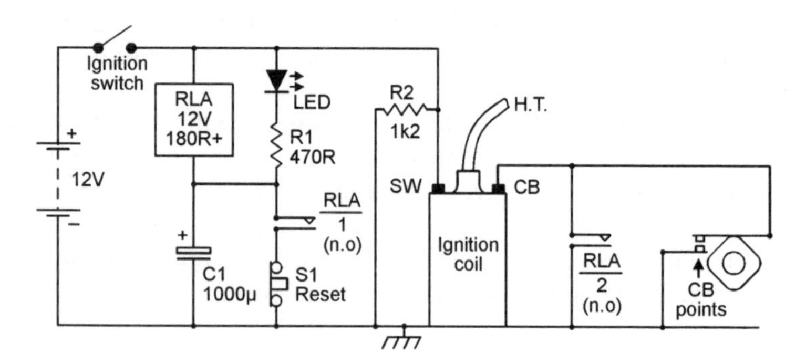

Figure 7.6 *Self-activating immobilizer circuit for 12V negative-ground vehicles*

A weakness of the *Figure 7.2* to *7.5* circuits is that they must all be turned on and off manually, and thus only give protection if the owner remembers to turn them on. By contrast, *Figure 7.6* shows an immobilizer that turns on automatically when an attempt is made to start the engine, but can be turned off by briefly operating a hidden push-button switch. A small 'reminder' light turns on when the engine is disabled by the immobilizer. This circuit thus gives a high level of protection, since it does not depend on the memory of its owner, and operates as follows.

The coil of relay RLA is shunted by series-connected R1 and a LED, which illuminates when RLA is on, and this combination is wired in series with 1000μF capacitor C1, which is shunted by series-connected n.o. relay contacts RLA/1 and n.c. push-button switch S1; this RLA–C1 combination is wired between the ignition coil's SW terminal and ground, and the relay's RLA/2 n.o. contacts are wired across the vehicle's CB points.

Normally, C1 is fully discharged; consequently, when the ignition switch is first closed a surge of current flows through RLA coil via C1, and the relay turns on, illuminating the LED. As the relay goes on, contacts RLA/1 close and lock the relay on via S1, and contacts RLA/2 close and short out the vehicle's CB points, thus immobilizing the engine. The relay stays on until S1 is briefly opened, at which point the relay unlatches and C1 charges up rapidly via the relay coil, and the relay and LED then turn off. As the relay turns off, it removes the short from the vehicles CB points, and the engine is able to operate in the normal way.

The relay used in the *Figure 7.6* circuit can be any 12V type with a coil resistance of at least 180R, and with two sets of n.o. contacts. The circuit shown is designed for use on vehicles with normal negative-ground 12V electrical systems; some older vehicles use 12V positive-ground systems, and on these simply reverse the polarities of C1 and the LED.

Anti-theft alarm circuits

It was explained earlier in this chapter that the most reliable vehicle anti-theft alarms are externally switched microswitch-activated or voltage-sensing types, and that on older vehicles these can be inexpensively turned on and off via a prominent key-switch (or a concealed toggle switch) fitted to the outside of the vehicle. *Figures 7.7* to *7.14* show practical examples of alarm systems of these types; each of these circuits can also act as an immobilizer, and (if required) operates the vehicle's horn and lights and immobilises the engine under the 'alarm' condition.

In the *Figure 7.7* to *7.12* circuits, microswitches that are built into the vehicle are used to trip a pair of self-latching relays when any of the car doors, hood or trunk are opened; these relays immobilize the engine and operate the horn and headlights either directly or via additional timing circuitry. Two suitable front-door microswitches are built into most vehicles as standard fittings and are used to operate the courtesy or dome lights; additional dome-light-operating microswitches, which are readily available from specialist 'vehicle security' retailers, can easily be fitted to the rear doors. Similar microswitches can be used as 'auxiliary' switches to protect the hood and trunk (bonnet and boot).

The operation of the *Figure 7.7* circuit is very simple. Normally, with the key-switch open, no voltage is fed to the relay network, so the alarm is off. Suppose, however, that the key-switch is closed. If any of the door switches close, current flows through both relays via D1, or if any of the auxiliary

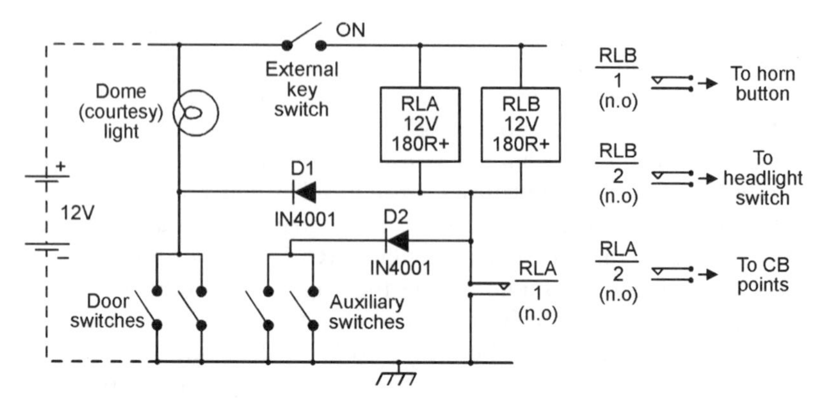

Figure 7.7 *Basic microswitch-activated anti-theft alarm/immobilizer for 12V negative-ground vehicles*

switches close, current flows through both relays via D2. In either case, both relays turn on. As RLA goes on, contacts RLA/1 close and lock both relays on, and contacts RLA/2 close and short out the vehicle's CB points, thus immobilizing the vehicle.

Simultaneously, contacts RLB/1 close and switch on the car horn (or a siren unit), giving an audible indication of the intrusion, and contacts RLB/2 close and switch on the headlights (or activate a headlight flasher unit), giving a visual identification of the violated vehicle. The horn and light (etc.) remain on until the key-switch is opened.

The *Figure 7.7* circuit is meant for use on vehicles fitted with normal 12V negative-ground electrical systems. The circuit can be modified for use on old vehicles fitted with 12V positive-ground systems by simply reversing the polarities of D1 and D2, as shown in *Figure 7.8*.

Obvious weaknesses of the simple *Figure 7.7* and *7.8* circuits are that, if they are activated and are not switched off manually within a reasonable space of time (typically 15 minutes), they will probably break local noise-control regulations, will probably damage the horn, and will eventually flatten the vehicle's battery. *Figure 7.9* shows a way of modifying the *Figure 7.7* circuit so that the horn and lights turn off automatically after about four minutes, but the immobilizer stays active until turned off via the key-switch, thus eliminating all of the above problems.

In *Figure 7.9*, RLA turns on and self-latches in the same way as in the *Figure 7.7* circuit, but as contacts RLA/1 close they connect the full battery voltage across the Q1–Q2 RLB-driving timer network, which has its timing controlled via C1–R1. At the moment that power is applied to this network, C1 is fully discharged and acts like a short circuit, so the base and collector of Q1 are effectively shorted together, thus driving RLB on via the Q1–Q2 Darlington emitter-follower and activating the horn and lights via contacts RLB/1 and RLB/2.

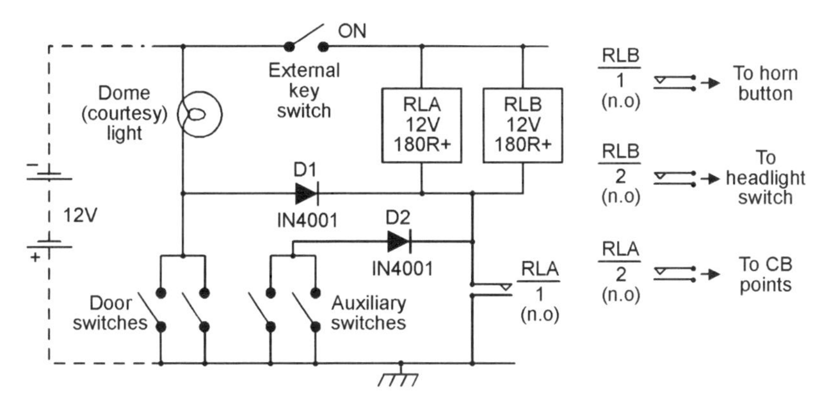

Figure 7.8 *Basic microswitch-activated anti-theft alarm/immobilizer for 12V positive-ground vehicles*

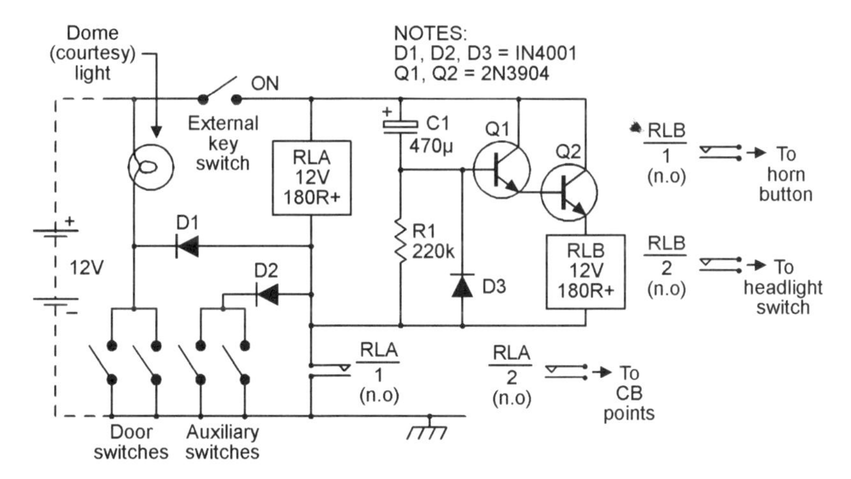

Figure 7.9 *Improved anti-theft alarm/immobilizer (for 12V negative-ground vehicles) has time-controlled outputs*

As soon as power is applied to the circuit, C1 starts to charge up via R1, and the voltage on RLB's coil starts to decay exponentially towards zero. After a delay of about four minutes this voltage falls so low that RLB (and thus the horn and lights) turn off. RLA remains on, however, until the system is turned off via the key-switch, and the vehicle thus remains immobilized via its CB points.

The *Figure 7.9* circuit is meant for use on 12V negative-ground vehicles. It can be modified for use on old 12V positive-ground vehicles by reversing the polarities of D1 and D2, and reversing the connections to the RLB-driving network, as shown in *Figure 7.10*.

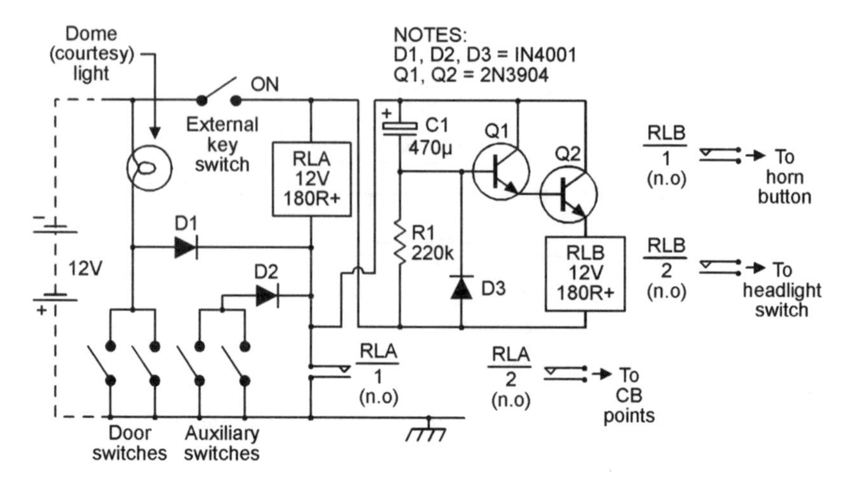

Figure 7.10 *Improved anti-theft alarm/immobilizer (for 12V positive-ground vehicles) has time-controlled outputs*

A practical snag with the above two circuits is that, since they give a 'monotone' form of horn operation, their owners are unlikely to be able to recognize the sound of their own vehicles, and will tend to check them whenever they hear any vehicle horn sound off. This snag is overcome in the

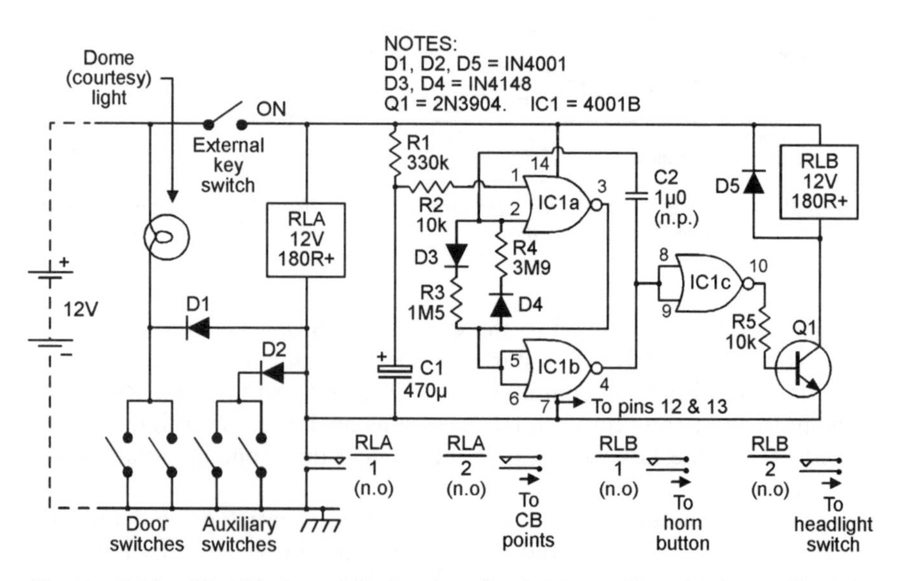

Figure 7.11 *Improved anti-theft alarm/immobilizer (for 12V negative-ground vehicles) has pulsed and time-controlled outputs*

circuit of *Figure 7.11*, which pulses the horn and lights on for 4 seconds and off for 1.5 seconds repeatedly for about four minutes under the alarm condition, thus producing a very distinctive warning signal.

The *Figure 7.11* circuit is similar to that of *Figure 7.9*, except that relay RLB is activated via a time-gated asymmetrical pulse generator formed from a 4001B CMOS IC and Q1. This generator is activated for about four minutes, via R1–C1, whenever RLA is turned on by a 'break-in' detector, and repeatedly pulses RLB (and thus the vehicle's lights and horn) on and off, for unequal periods, for the duration of this 4-minute period. During this 'pulsing' period, the 'off' time of RLB is controlled by C2–D3–R3 and approximates 1.5 seconds, and the 'on' time is controlled by C2–D4–R4 and approximates 4 seconds. Note that C2 is a non-polarized (n.p.) capacitor.

The *Figure 7.11* circuit is designed for use on vehicles fitted with 12V negative-ground electrical systems; it can be modified for use on old 12V positive-ground vehicles by reversing the polarities of D1 and D2 and the supply connections to the RLB-driving network, as in *Figure 7.12*.

Finally, to complete this look at anti-theft alarm circuits, *Figure 7.13* shows the practical circuit of a simple voltage-sensing type of alarm unit that can be used in place of the basic RLA-driving network used in the *Figure 7.7* to *7.12* circuits. Circuit operation relies on the fact that (when the vehicle's engine is not operating) a small but sharp drop occurs in the vehicle's battery voltage whenever a courtesy light is automatically turned on by the opening of a front door, or when the ignition is switched on. This sudden drop in voltage is detected and made to operate RLA. The system has the advantage that its

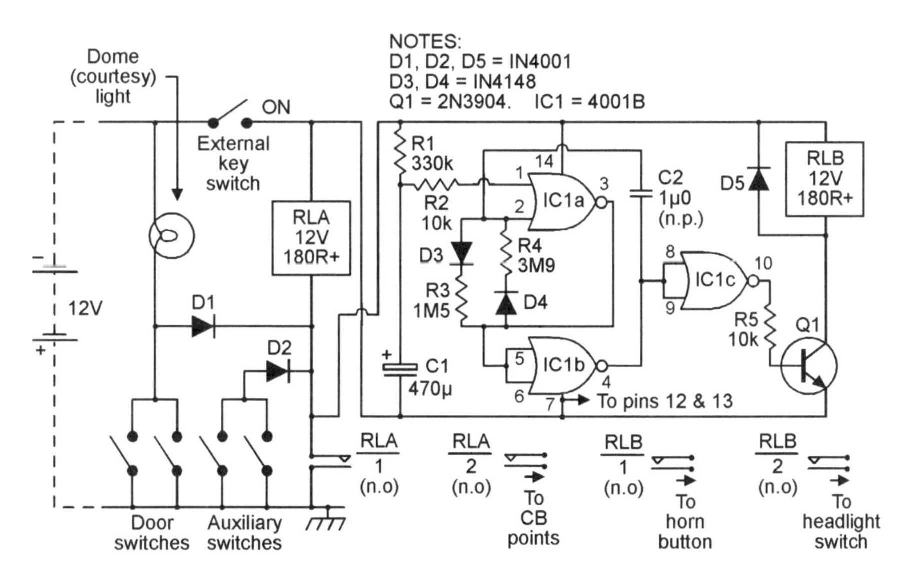

Figure 7.12 *Improved anti-theft alarm/immobilizer (for 12V positive-ground vehicles) has pulsed and time-controlled outputs*

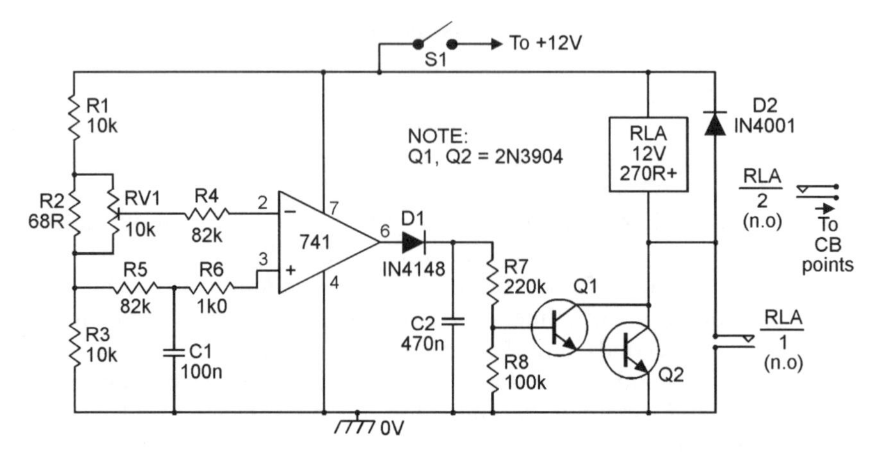

Figure 7.13 *Basic voltage-sensing alarm circuit for 12V positive-ground vehicles*

'alarm' signals are derived directly from the vehicle's battery, rather than via various microswitches, but it is not quite as reliable as a conventional microswitch-activated alarm system.

The operation of the *Figure 7.13* circuit – which is connected across the vehicle's battery via S1 – is fairly simple. Here, potential divider R1–R2–R3 is wired across the circuit's supply lines, and the output of this divider is fed, via RV1–R4, to the inverting (pin-2) input terminal of the 741 op-amp, which is wired in the open-loop mode, but is taken to the non-inverting (pin-3) input via a simple (R5–C1–R6) time-delay 'memory' network. A small offset voltage can be applied between the two input terminals via RV1.

Suppose then that the RV1 offset control is adjusted so that the pin-2 voltage is fractionally higher than that of pin-3 under normal 'steady voltage' conditions, and that under this condition the output of the op-amp is driven to negative saturation. If now a small but abrupt fall occurs in the supply voltage, this fall is transferred immediately to pin-2 of the op-amp, but does not immediately reach pin-3 because of the time-delay or memory action of C1. Consequently, pin-2 briefly goes negative relative to pin-3, and as it does so the output of the op-amp is driven briefly to positive saturation, thus giving a positive output pulse. This pulse is used to charge C2 via D1, and C2 drives Q1–Q2 and the relay on via R7. As the relay goes on, contacts RLA/1 close and cause the relay to self-latch, and contacts RLA/2 close and immobilize the vehicle via its CB points. Note that this circuit responds only to sudden drops in voltage, and is not influenced by stable absolute values of battery voltage.

The *Figure 7.13* circuit is intended for use on 12V negative-ground vehicles, and can be used directly in place of the RLA network in any of the *Figure 7.7*, *7.9*, or *7.11* circuits. The circuit can be modified for use on 12V

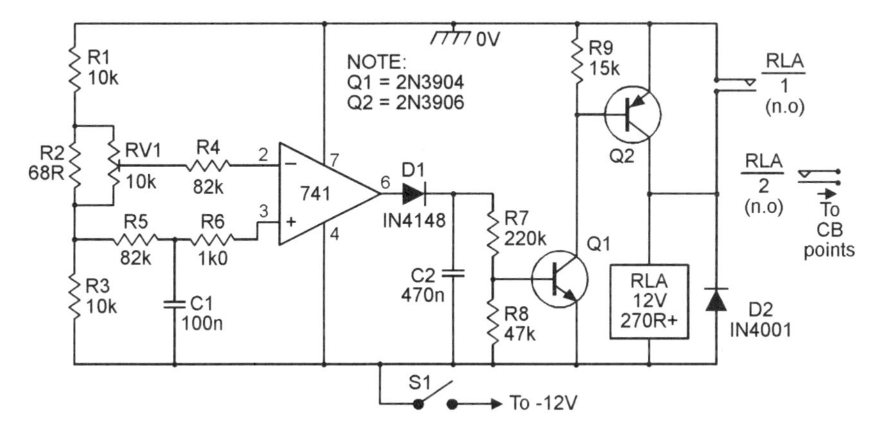

Figure 7.14 *Basic voltage-sensing alarm circuit for 12V negative-ground vehicles*

positive-ground vehicles by using the connections shown in *Figure 7.14*, and can then be used directly in place of the RLA network in any of the *Figure 7.8*, *7.10* or *7.12* circuits.

When installing the *Figure 7.13* or *7.14* circuit in a vehicle, RV1 must be carefully adjusted so that the alarm turns on reliably when the courtesy (dome) light goes on, without being excessively sensitive to the small shifts that occur in the battery voltage due to its chemical action. To find the correct setting of RV1, temporarily disable self-latching contacts RLA/1, and temporarily replace the courtesy lamp with one having roughly half its normal current rating. Now adjust RV1 just past the point where RLA fails to activate when the lamp goes on, and then turn RV1 back a fraction, so that RLA is just activated by the courtesy light. Now refit the original courtesy light, and recheck the action. If reliable action is obtained, re-enable the RLA/1 contacts.

Installing anti-theft alarms

The anti-theft alarms described in this chapter are all designed to be turned on and off via an externally mounted switch, which may take the form of a carefully hidden toggle switch or a prominently mounted key-switch. In either case, the switch must be mounted so that neither it or its wiring is vulnerable to damage by weather, dirt or potential car thieves.

Once the alarm's external on/off switch has been fitted, the next installation job is to fit suitable microswitches to activate the system. Two suitable switches are already fitted to most vehicles, and are used to operate the dome or courtesy light. Additional switches must be fitted to the rear doors, and must also be fitted to the trunk and hood if full anti-theft protection is to be obtained. Note that if your vehicle is fitted with a voltage-sensing type of alarm system, these microswitches must be used to switch a normal filament lamp; the higher the load current used, the more reliable will be the operation of the alarm circuit; the microswitches can all be wired in parallel and used to drive a single load. When installation is complete, give your system a complete functional check, taking care not to annoy your neighbours in the process.

Ice-hazard alarms

Ice-hazard alarms activate when the vehicle's ignition is turned on and the air temperature several inches above the road surface is at or below the 0°C freezing point of water. The alarms thus indicate a risk of meeting ice under actual driving conditions. Two useful ice-hazard alarms are shown in this section, and can easily be fitted to most automobiles. Both units use a low-cost n.t.c. thermistor – mounted outside of the vehicle, near its front and several inches above the road surface – as a thermal sensor that gives a good indication of the actual road temperature.

The first circuit, shown in *Figure 7.15*, connects to the vehicle's 12V supply via its ignition switch, and turns on relay RLA under the 'ice-hazard' condition; the RLA/1 contacts can be used to activate any desired type of external alarm device. In this circuit, the 3140 op-amp is used as a voltage comparator, in which a pre-settable reference voltage is applied to its pin-3 input via RV1, a temperature-sensitive variable voltage is applied to pin-2 from the R1–TH1 potential divider, and the action is such that the pin-6 output of the op-amp switches low and activates Q1 and RLA if the variable voltage exceeds the reference voltage. In practice, the output voltage of the R1–TH1 divider rises as the TH1 temperature falls, and – if the circuit is correctly set up – trips relay RLA at a TH1 temperature of precisely 0°C.

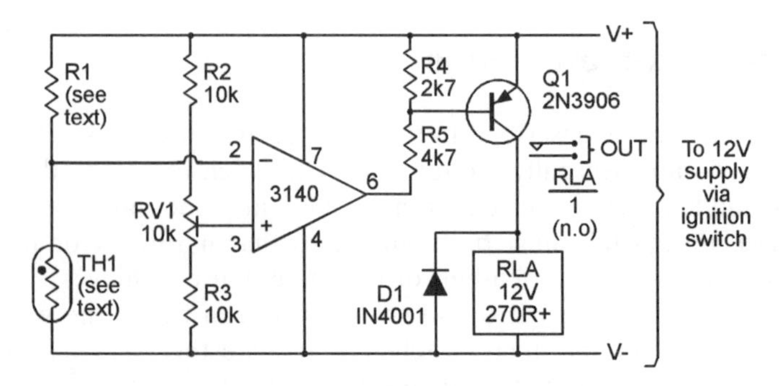

Figure 7.15 *Relay-output ice-hazard alarm*

This circuit's thermistor can be any n.t.c. disc or bead type that has a resistance in the range 1k5 to 5k0 at 25°C (the normally-specified 'nominal resistance value' temperature). Typically, the TH1 resistance at 0°C is about treble the 25°C value, and in this circuit the R1 value should roughly equal TH1's '0°C' value; thus, if TH1 has a nominal 25°C value of 1k5, R1 needs a value of about 4k5. Note, however, that the precise R1 value is not critical, since balance-control RV1 can compensate for errors in the range –50% to +100% in the R1 value. The actual methods of mounting TH1 and setting RV1 are described later in this section.

In most practical ice-hazard alarm applications the hazard condition may continue intermittently during several hours of driving, and it is thus best to use a flashing-LED type of 'hazard alarm' indicator, rather than a continuously-active audible-warning type. *Figure 7.16* shows the *Figure 7.15* circuit modified to give a flashing-LED alarm output directly, rather than via a relay. In this case Q1 is wired as an emitter-follower, and drives on D1 – a low-cost flashing LED device – when the op-amp output switches low. Note that R4–ZD1 and Q1's base-emitter volt-drop limit D1's maximum applied voltage to a safe value of 9.4 volts.

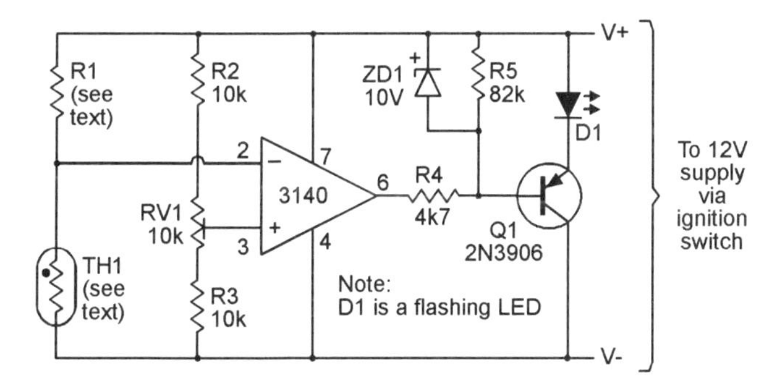

Figure 7.16 *Ice-hazard alarm with flashing-LED output*

When building either of these ice-hazard alarms, note that thermistor TH1 must be mounted in a small 'head' that is fixed to the lower front of the vehicle and connected to the main alarm-unit via twin flex. To make the thermistor head, solder the thermistor to a small tag-board and solder its leads to the twin flex. Coat the whole assembly with waterproof varnish, so that moisture will not affect its apparent resistance, then mount it in a small plastic or metal box and fix it to the lower front of the vehicle. Before fixing the head in place, however, calibrate the alarm system as follows:

Immerse the head in a small container filled with a water and ice mixture. Use a thermometer to measure the temperature of the mixture, and add ice until a steady reading of 0°C is obtained. Now adjust RV1 so that the alarm (RLA or the flashing LED) just turns on; raise the temperature slightly, and check that the alarm turns off again. If satisfactory, the head and the alarm system can now be fixed to the vehicle.

Low-fuel-level/engine-overheat alarms

Two of the most annoying things that can happen to a car driver are to run out of fuel, or to have the engine suddenly fail due to severe overheating, when each of these conditions could easily have been avoided if the driver had noticed the advance-warnings indicated by the vehicle's fuel gauge or the engine's water-temperature gauge readings. Both of these events can often be avoided by fitting the vehicle with simple circuits that monitor the standard fuel or temperature gauges and activate a flashing-LED alarm under potential 'danger' conditions. Two such circuits are shown in this section; both use the same flashing-LED type of output stage as the *Figure 7.16* circuit, but can be made to give relay outputs by replacing these with the type of output stage used in the *Figure 7.15* circuit.

Most reasonably-modern vehicles are fitted with an analogue fuel-level gauge circuit that takes the basic form shown in *Figure 7.17*. Here, the gauge is actually a hot-wire (bimetal) current meter that is wired in series with a 'sender' unit that is mounted in the fuel tank and consists of a float-driven rheostat that presents a high resistance (and a low current-meter reading) when the tank is empty, and a low resistance (and a high current-meter reading) when the tank is full. Note that the output voltage of the sender rises as the fuel level falls, and a low-fuel-level alarm can thus take the basic form of a simple over-voltage alarm. A suitable 'flashing-LED' alarm circuit is shown in *Figure 7.18*.

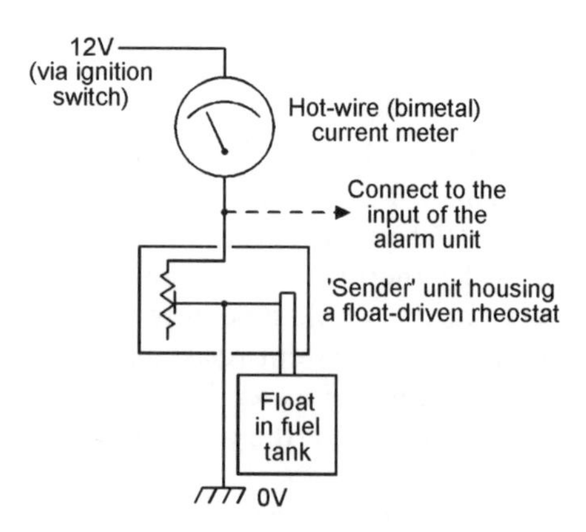

Figure 7.17 *Basic circuit of a typical fuel-level gauge*

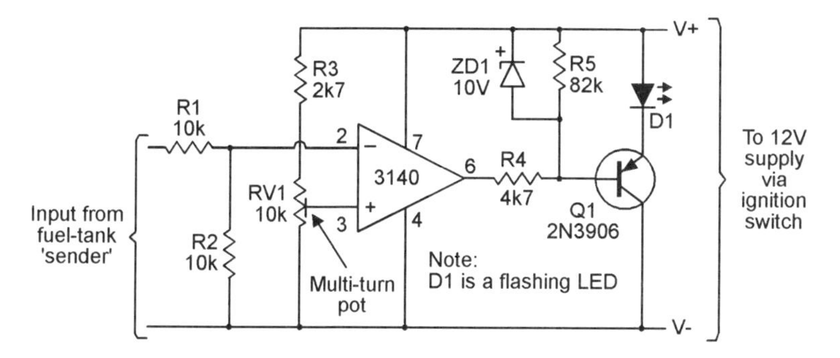

Figure 7.18 *Low-fuel-level alarm*

Before starting to build the *Figure 7.18* circuit, check that you can gain easy access to the tank-mounted sender unit or the rear of the fuel gauge, and that the sender's output voltage varies in the way described above. In the *Figure 7.18* circuit, the 3140 op-amp is wired as a voltage comparator, with a pre-set fraction of the supply voltage applied to pin-3 via multi-turn pot RV1, and with half of the sender's output voltage applied to pin-2 via the R1–R2 potential divider. The op-amp's output goes low and activates the flashing-LED (via Q1) when the pin-2 voltage rises above the pre-set pin-3 value.

To set up the *Figure 7.18* circuit, wait until the fuel falls to the required 'danger' level, then connect the unit to the vehicle's supply via the ignition switch and trim RV1 so that the flashing-LED just turns on. Check that the flashing-LED turns off when the fuel level is increased by a modest amount.

Most reasonably-modern vehicles are fitted with an analogue engine-temperature gauge circuit that takes the basic form shown in *Figure 7.19*. Here, the gauge is a hot-wire current meter that is wired in series with a

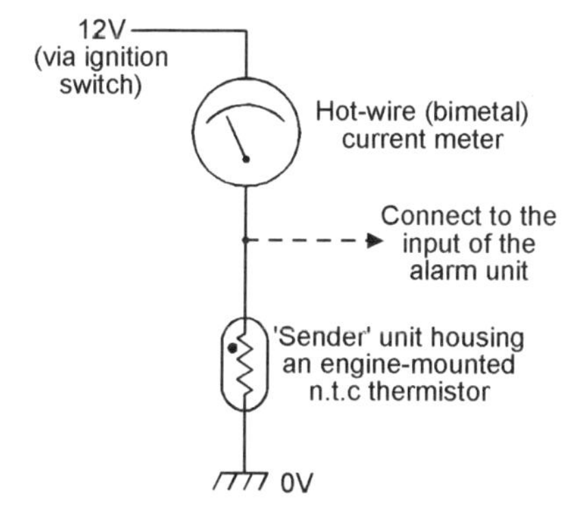

Figure 7.19 *Basic circuit of a typical engine-temperature gauge*

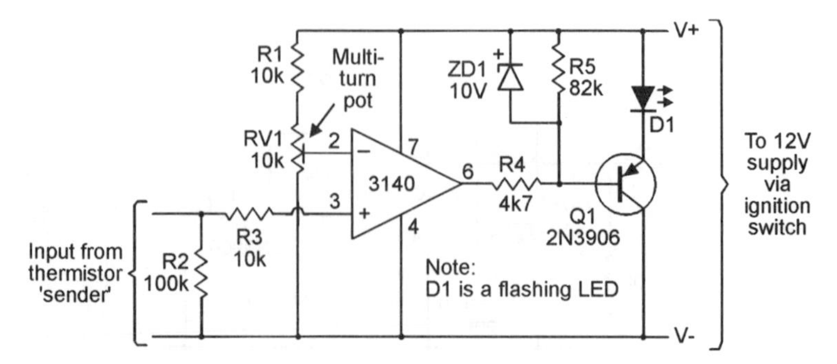

Figure 7.20 *Engine over-heat alarm*

'sender' that is simply an engine- or radiator-mounted n.t.c. thermistor and presents a high resistance (and a low current-meter reading) at low temperatures, and a low resistance (and a high current-meter reading) when the engine is at or above its normal running temperature (about 105°C *absolute maximum* in modern engines). Note that the output voltage of the sender falls as the engine temperature rises, and an engine over-heat alarm can thus take the basic form of a simple under-voltage alarm. A suitable 'flashing-LED' alarm circuit is shown in *Figure 7.20*.

Before starting to build the *Figure 7.20* circuit, gain access to the sender unit and check that its output voltage varies in the way described above. In the *Figure 7.20* circuit, the 3140 op-amp is wired as a voltage comparator, with a pre-set fraction of the supply voltage applied to pin-2 via multi-turn pot RV1, and with the sender's output voltage applied to pin-3 via R2–R3. The op-amp's output goes low and activates the flashing-LED (via Q1) when the pin-3 voltage falls below the pre-set pin-3 value.

To set up the *Figure 7.20* circuit, connect it to the vehicle's supply via the ignition switch, temporarily disable the cooling fan, run the engine up to a fairly high 'X' temperature and trim RV1 so that the flashing-LED just turns on, then stop the engine and let it cool down. Now restart the engine and check that the flashing-LED remains off until 'X'°C is reached. If all is well, turn the LED off via RV1, run the engine up to its red-line 'overheat' temperature, then reset RV1 so that the LED turns on. Finally, stop the engine and re-enable the cooling fan.

Lights-are-on alarms

Most modern automobiles are fitted – as standard equipment – with a lights-are-on reminder unit that emits a low-power audible alarm signal that warns the driver, as he/she open the driver's door, if the car lights have been left on. 'Lights-are-on' alarm units of this basic type are easy to build and are

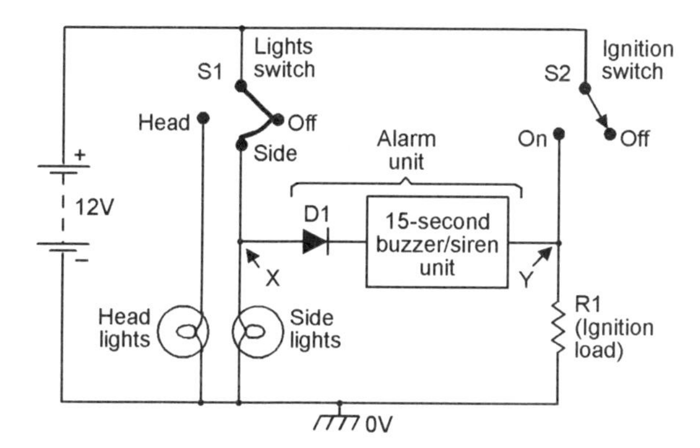

Figure 7.21 *Basic Type-1 'lights-are-on' alarm circuit for 12V negative-ground vehicles*

fairly easy to add to most older types of vehicle, and come in two basic types. Type-1 is for use in very old vehicles that are *not* fitted with a courtesy light that is switched automatically via a conventional door-activated microswitch, and Type-2 is for use in vehicles that are fitted with such a courtesy light system.

Figure 7.21 illustrates the basic operating principles of a Type-1 lights-are-on alarm that is added to a vehicle with a 12V negative-ground electrical system. Here, the actual alarm consists of a low-power 15-second buzzer or siren unit that has its supply lines wired in series with diode D1 so that current can only flow through the unit (and thus activate the buzzer/siren) when point-X is many-volts positive to point-Y, and in the diagram this situation only occurs when 'lights' switch S1 is in the ON (Side or Head) position and 'ignition' switch S2 is turned OFF. Thus, if the lights are left on, the alarm unit sounds-off as soon as the ignition is switched off, but mutes automatically after 15 seconds.

Figure 7.22 illustrates the basic operating principles of a Type-1 lights-are-on alarm that is added to a vehicle with a 12V positive-ground electrical system. Here, the positions of the 'X' and 'Y' points are simply transposed, so that the alarm again activates only when 'light' switch S1 is ON and 'ignition' switch S2 is OFF.

Figure 7.23 shows a simple but practical version of a Type-1 'lights-are-on' alarm unit. Here, Q1–Q2 are wired as a modified complementary astable multivibrator that has its power supplied via D1 and uses a 64R low-power speaker as the collector load of Q2, which is a 2N3704 type and has a peak collector current rating of 800mA. The astable's action is such that it generates a loud and fairly high alarm tone when power is initially applied, but the volume and frequency then decay steadily down to zero over a period of about 15 seconds. The decay time is controlled by R3–C2, and the initial

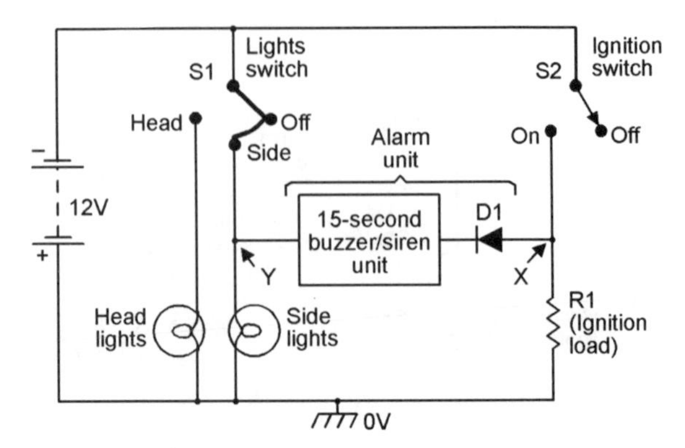

Figure 7.22 *Basic Type-1 'lights-are-on' alarm circuit for 12V positive-ground vehicles*

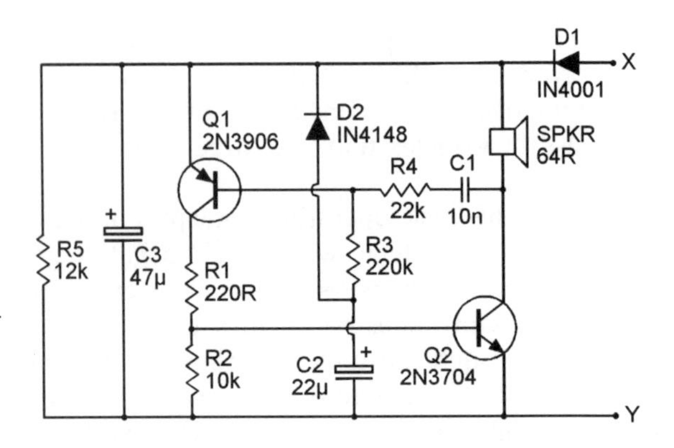

Figure 7.23 *Practical Type-1 'lights-are-on' alarm unit*

frequency is controlled by R3–R4 and C1. To use the unit, simply connect its 'X' and 'Y' points in the appropriate way shown in *Figure 7.21* or *7.22*.

Figure 7.24 and *7.25* illustrate the basic operating principles of Type-2 lights-are-on alarms that are added to vehicles with 12V negative-ground or positive-ground electrical systems. The Type-2 system is similar to the Type-1 system, except that the alarm unit is wired between 'lights' switch S1 and dome-light-activating door-switch S2, so that the alarm only sounds if the lights are on when the driver's door is open.

The above system is used on most modern vehicles, in which the dome light uses separate circuits for its 'courtesy' and 'reading' lights. A snag with most older vehicles is that they use a single light-bulb to perform both of these functions, with the 'reading' switch wired in parallel with S2, as shown by S3 in *Figures 7.24* and *7.25*, and as a consequence the alarm can be

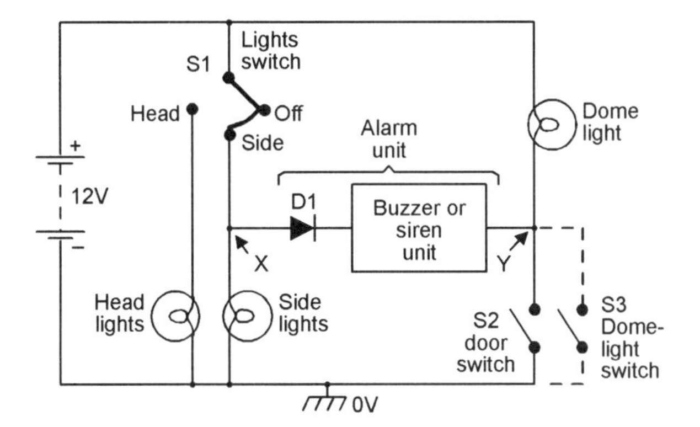

Figure 7.24 *Basic Type-2 'lights-are-on' alarm circuit for 12V negative-ground vehicles*

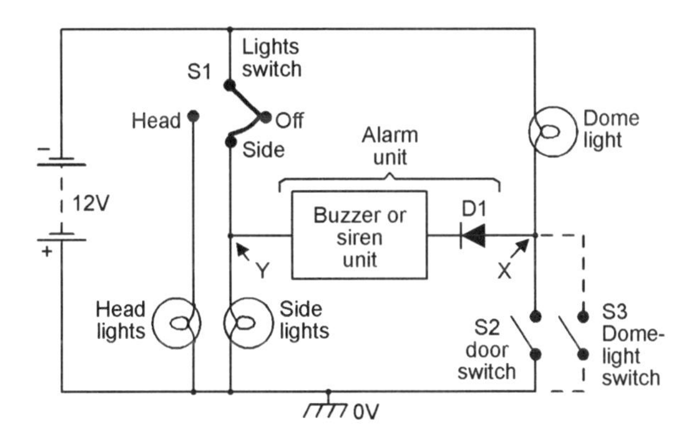

Figure 7.25 *Basic Type-2 'lights-are-on' alarm circuit for 12V positive-ground vehicles*

activated by either of these switches. The easiest way around this snag is to use an auto-turn-off alarm unit that, like the Type-1 unit, only sounds for a dozen or so seconds when activated. The alarm unit can thus take the simple 'monotone' form shown in the *Figure 7.23* circuit, or can take the more attractive 'warble-tone' form shown in the *Figure 7.26* circuit.

The *Figure 7.26* circuit is based on the warble-tone circuits that are shown in *Figures 2.17* and *2.23* and are fully described in Chapter 2. In brief, however, IC1a–IC1b and R4–C2 are wired as a gated l.f. (low-frequency) astable and IC1c–IC1d and R5–C3 form a gated h.f. (high-frequency) astable. Both astables are gated via the R2–C1 time-constant network, and are on when C1's output is below (roughly) half-supply volts and off when C1's output is above half-supply volts. When the astables are gated on, the l.f.

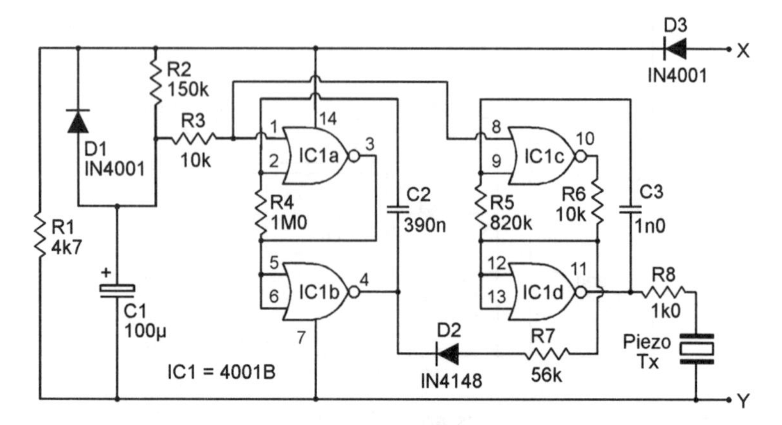

Figure 7.26 *Practical warble-tone Type-2 'lights-are-on' alarm unit*

astable frequency-modulates the h.f. astable (via D2–R7–R6), which has its output fed to piezo output transducer Tx via R8.

When power is connected to the *Figure 7.26* circuit, C1 is initially fully discharged, so both astables are gated on and generate a warble-tone sound in the piezo Tx, but C1 then charges up via R2 and after a delay of about 15 seconds (determined by R2–C1) gates both astables off, silencing the piezo Tx. When the unit's power connections are removed, C1 rapidly (in a second or so) discharges via D1–R1, and the unit is then ready to repeat its operating sequence when power is connected again.

A headlight time-delay switch

Headlight time-delay switches are simple add-on units that, when activated by a push switch, turn a car's headlights on for just a few minutes, during which the owner can leave and lock the vehicle, walk along a drive or pathway that is well lit by the headlights, and then enter the security of a safe building before the lights automatically turn off again. *Figure 7.27* shows a practical transistor version of such a circuit.

In *Figure 7.27*, Q1 is an emitter follower that uses R3–R4 as its emitter load but has 'bootstrapped' resistor R2 wired between its base and emitter, so that R2's impedance – as seen from Q1's base – is many times greater than its dc resistance value. C1 and Q1's input impedance (which equals the parallel values of R2's impedance and Q1's base impedance) form a C–R time-constant network. The action of Q1 and this network is such that, when power is first connected to the circuit by briefly closing push-button switch S1, C1 is fully discharged and thus pulls Q1's base and emitter up to almost the full positive supply voltage, thus driving relay-activating common-emitter

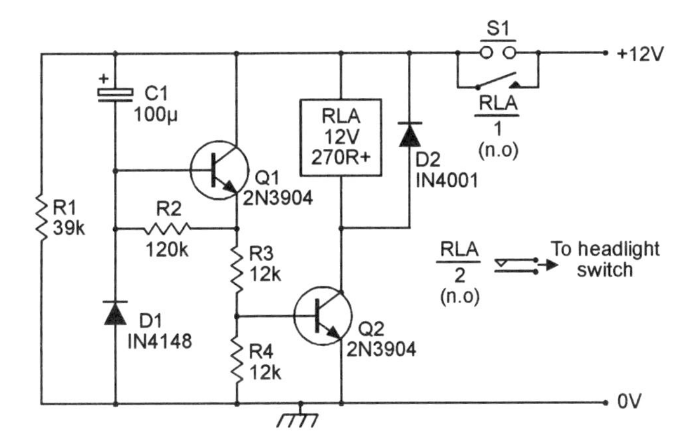

Figure 7.27 *A headlight time-delay switch circuit*

amplifier Q2 on via R3; as relay RLA turns on, contacts RLA/1 close and lock the supply on, and contacts RLA/2 close and turn the vehicle's headlights on.

As soon as power is connected to the circuit, C1 starts to charge up via the very high parallel impedances of R2 and Q2's base, and Q2's emitter voltage (and the current fed into Q2's base via R3) starts to decay exponentially until, after a delay of about two minutes, these values fall so low that RLA turns off, thus breaking the supply connection as RLA/1 opens and turning off the headlights as RLA/2 opens. C1 discharges rapidly via R1–D1 when the circuit's power connection is broken.

A rear-screen heater timer

Automobile rear-screen heaters draw typical operating currents of about 15 amps and thus place a heavy strain on the vehicle's electrical generating system. If the heater is inadvertently left on for long periods, this strain can greatly reduce the generator's reliability and working life. Consequently, most modern automobiles are fitted with push-button operated rear-screen heater controllers that turn the heater off automatically after an operating period of about 15 minutes. Reliability-enhancing units of this basic type can easily be added to older vehicles, and – to conclude this chapter – *Figure 7.28* shows a practical circuit of this type.

The *Figure 7.28* circuit incorporates a simple voltage regulator built around transistor Q1, a stable 16-minute timer built around IC1 and IC2, and relays RLA (which controls the circuit's semi-latching function) and RLB (which controls the rear-screen heater's power feed), and derives its power feed via the vehicle's ignition switch. The complete circuit operates as follows.

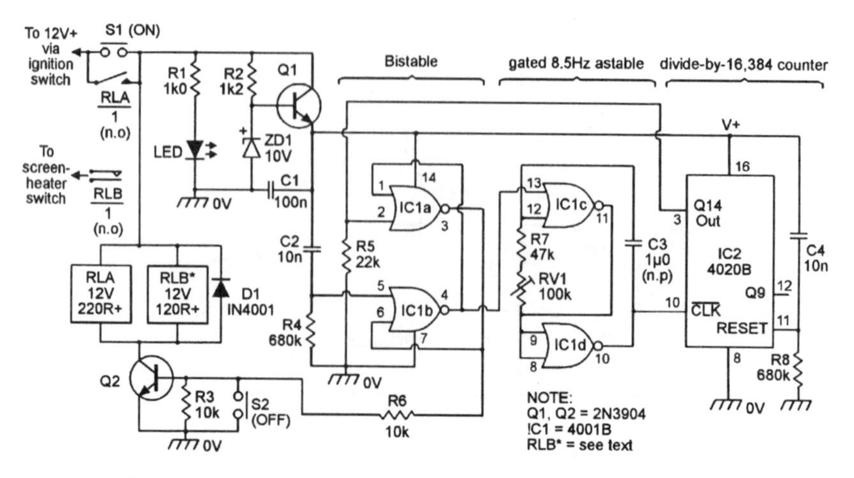

Figure 7.28 *A rear-screen heater timer circuit*

The *Figure 7.28* circuit is turned on by briefly closing push-button switch S1. As S1 closes, the LED illuminates, a stable 9.4V supply is applied to the IC1–IC2 timer circuit via Q1, and 'reset' pulses are fed to the IC1a–IC1b bistable (via C2–R4) and the IC2 counter (via C4–R8). As the bistable resets, its pin-3 output flips high and turns on RLA and RLB via Q2; as the relays turn on, the RLA/1 contacts close, shunting S1 and locking-on the circuit's supply connection, and RLB/1 contacts close and connect power to the rear-screen heater. Simultaneously, as the IC1a–IC1b bistable resets, its pin-4 output flips low and turns on the IC1c–IC1d gated astable, which starts feeding clock pulses into pin-10 of the IC2 counter at an 8.5Hz rate. Sixteen minutes later, on arrival of the astable's 8192nd clock pulse, the pin-3 output of IC2 flips high and changes the state of the IC1a-IC1b bistable, which simultaneously gates off the astable and removes Q2's base drive, thus turning off both relays; as RLA turns off it removes the timer's supply connection, and as RLB turns off it removes power from the rear-screen heater.

Thus, when the vehicle's ignition is turned on, the *Figure 7.28* circuit connects power to the rear-screen heater as soon as S1 is closed, but removes it again automatically after 16-minutes; if desired, the timing period can be ended prematurely by briefly closing S2, thus turning off both relays. The circuit's timing period can, if you wish, be set to precisely 16 minutes by connecting a LED and 2k7 series resistor between pins 12 and 8 (0V) of IC2 and trimming RV1 so that the LED operates with precise 30-second on and off periods. When building the circuit, note that RLA's contacts have to pass maximum currents of less than 200mA, and RLA can thus be almost any general-purpose relay, but that RLB's contacts have to pass the full operating current of the rear-screen heater (typically 15A), and RLB must thus be a dedicated heavy-duty 'automobile' relay.

8

Miscellaneous security circuits

Chapters 2 to 7 of this book have each looked at a specific class or type of electronic security circuit. This concluding chapter looks at a miscellaneous collection of security circuits that can be used in the home or in commerce or industry, but which do not fit into any of the specific classes of circuit described in earlier chapters.

The circuits described in this chapter include ones that are activated by the presence of a liquid, steam, or gas, by sound, by the failure of AC power supplies, by the close or near proximity of a person or object, by a human touch, or by the breaking of an ultrasonic beam. The chapter also describes two simple 'eavesdropping' devices, and describes basic ways of using CCTV cameras in security applications.

Liquid- and steam-activated circuits

Basic principles and circuits

Liquid- and steam-activated circuits have several practical applications in the home and in industry. Liquid-activated circuits can be made to sound an alarm or activate a safety mechanism when the water in a bath or cistern or the liquid in a tank or vat reaches or exceeds (or falls below) a pre-set level, or when rain water touches a pair of contacts, or when flooding occurs in a cellar or basement, or when an impact wave is generated as a person or object falls into a swimming pool or tank, etc. Steam-activated circuits can be made to sound an alarm or activate a safety mechanism when high-pressure steam escapes from a valve or fractured pipe, or when steam emerges from the spout of a kettle or container as the liquid reaches its boiling point.

Impure water (including tap water, sea water, and most rain water and steam) and many other liquids have a fairly low electrical resistance, but

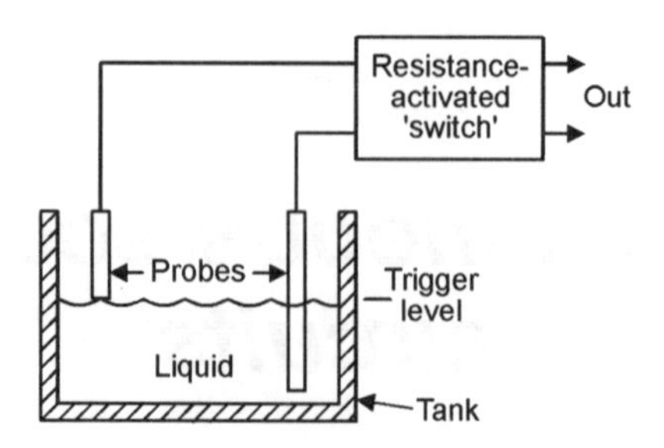

Figure 8.1 *Basic 'electronic' way of detecting conductive liquids via a pair of metal probes*

normal air has an ultra-high electrical resistance. Consequently, one of the simplest *electronic* ways of detecting the presence or absence of conductive liquids (or vapours) is to use a pair of metal probes as sensors and to connect their outputs to a resistance-activated 'switch' circuit in the basic way shown in *Figure 8.1*.

Here, when the liquid is in contact with both probes simultaneously the probe-to-probe resistance is relatively low, and under this condition the output voltage of the resistance-activated switch is also low, but when the liquid is not in contact with both probes at the same time their probe-to-probe resistance is very high, and under this condition the output voltage of the switch is high. The circuit's output can thus be used to activate an alarm or other device when the liquid (or vapour) is present or absent, or is above or below a pre-set level.

In practice, the resistance appearing across the probes under the 'contact' condition depends on the type of medium that is being detected. In the case of rain or tap water it may typically be in the range 1k0 to 10k when the probes are 10mm apart, but in the case of steam or many oils the resistance may be several megohms or greater. On some applications, one of the two metal probes may actually be the metal container or tank that holds the conductive liquid.

Note that the above 'liquid-detection' technique is not suitable for use with highly volatile, corrosive, or highly resistive liquids. In such cases, the presence\absence of the liquid may best be detected by using an electro-mechanical method, such as that shown in *Figure 8.2*. Here, the liquid is contained in a sealed tank, and its level is sensed by a float that is anchored to a pivoted lever that drives a rod that passes out of the top of the tank via a close-fitting tube. The rod thus rises and falls in sympathy with the liquid level, and activates an external microswitch when the level goes above or below some pre-set limit.

Figures 8.3 and *8.4* show two practical examples of simple non-latching liquid-activated circuits that operate a relay or a self-interrupting alarm bell

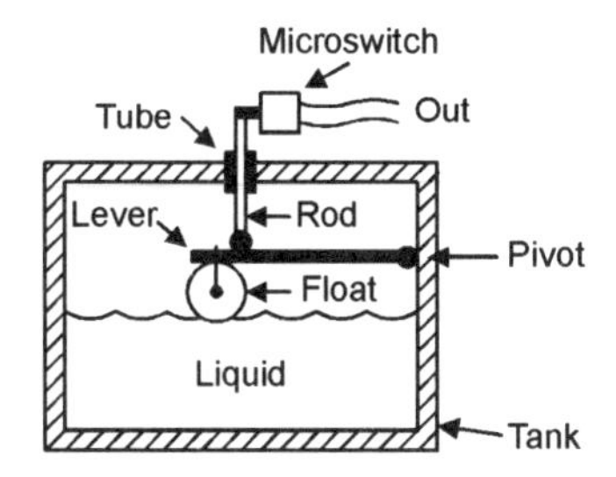

Figure 8.2 *Basic 'electromechanical' way of detecting the critical level of a liquid*

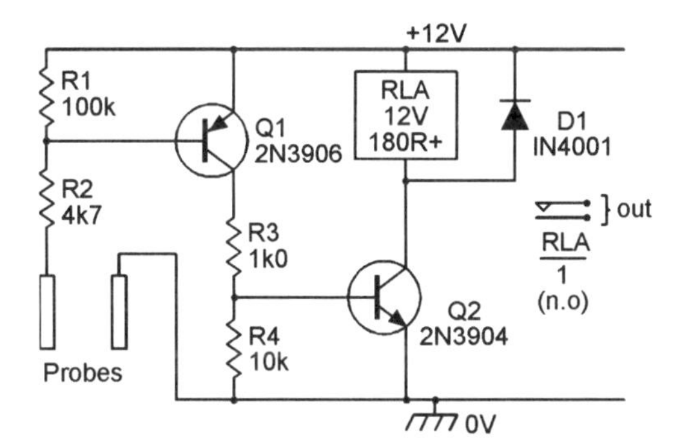

Figure 8.3 *Simple relay-output 'liquid-activated' circuit operates when less than 500k is applied across the probes*

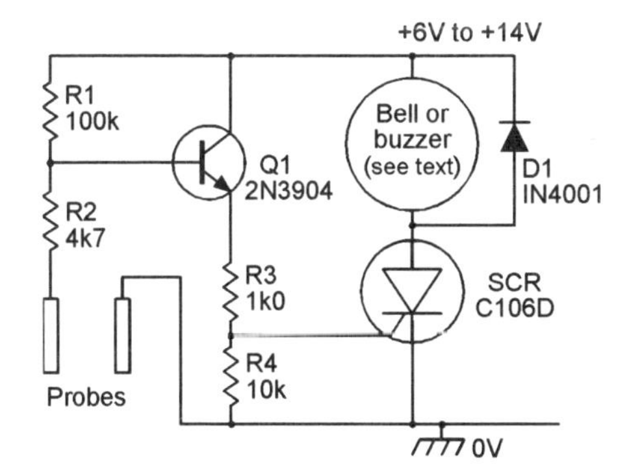

Figure 8.4 *Simple bell/buzzer-output 'liquid-activated' circuit operates when less than 500k is applied across the probes*

or buzzer when a liquid with a resistance of less than about 500k contacts both probes simultaneously, e.g. when the water in a bath or cistern reaches a certain level.

The *Figure 8.3* circuit uses a 12V supply and activates a 12V relay under the water 'contact' condition, and can activate any type of external electrical device via the RLA/1 contacts. When the probes are open circuit, Q1 and Q2 are both cut off and the circuit consumes a standby current of less than 1μA, but when a resistance of less than about 500k is applied across the probes sufficient current flows in Q1's base to drive Q1–Q2 and the relay fully on.

The *Figure 8.4* circuit activates a self-interrupting bell or buzzer under the water contact condition; the bell/buzzer must pass an operating current of less than 2A, and the supply voltage must be about 1.5V greater than the alarm device's nominal operating voltage. When this circuit's probes are open circuit, Q1 and the SCR are both cut off and the circuit consumes a standby current of less than 1μA, but when a resistance of less than about 500k is applied across the probes sufficient current flows in Q1's base to drive Q1 and the SCR fully on, and the alarm device activates.

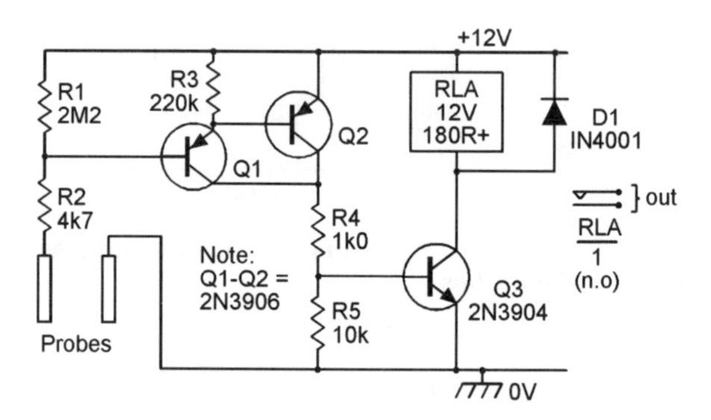

Figure 8.5 *Sensitive relay-output 'liquid-activated' circuit operates when less than 20M is applied across the probes*

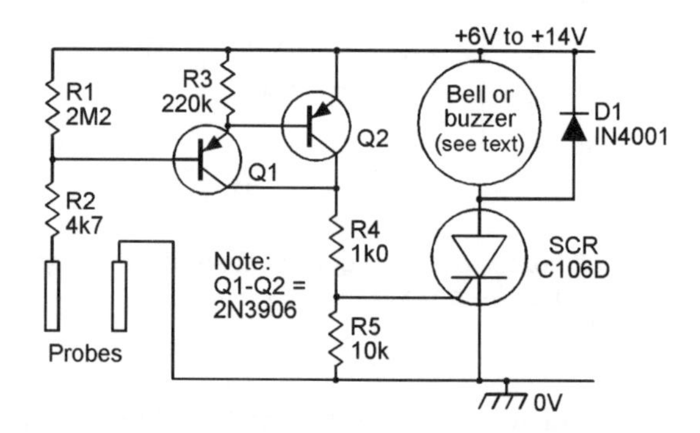

Figure 8.6 *Sensitive bell/buzzer-output 'liquid-activated' circuit operates when less than 20M is applied across the probes*

Note that the sensitivities of the *Figure 8.3* and *8.4* circuits can be reduced by simply reducing the value of R1; in the *Figure 8.3* circuit, for example, the *maximum* resistive sensitivity is roughly 18 × R1, and falls below 180k when R1 has a value of 10k, and below 60k at an R1 value of 3k3. Conversely, the sensitivity can be greatly increased by raising the R1 value and using a super-alpha-connected pair of transistors in place of Q1, and *Figures 8.5* and *8.6* show how the above two circuits can be modified in this way, so that they can be activated by probe resistances of up to 20M, e.g. by steam or high-resistance liquids.

LM1830 IC circuits

When in use, liquid-level detector circuits of the simple types shown in *Figures 8.3* to *8.6* pass a small dc current through the liquid under test. In theory, this dc current can result in an electroplating action in which metal slowly migrates from one probe to the other, eventually degrading the 'source' probe. This problem does not occur if an ac test current is used, and a dedicated 'fluid-level detector' IC that uses this technique is widely available, at modest cost; the device is manufactured by National Semiconductor and is known as the LM1830; *Figure 8.7* shows the outline and simplified internal circuit of the IC.

The LM1830 can be used to detect and indicate the presence or absence of water or any other liquid that presents a resistance of less than 100k between its pin-10 (detector input) and pin-11 (GND) 'probe' points. The IC houses an oscillator (which gives ac drive to the water-detecting metal probe), a 13k reference resistor, a balance detector, and (available on pin-

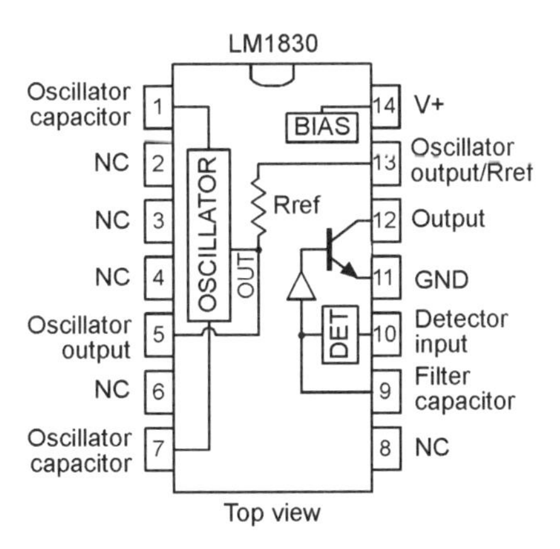

Figure 8.7 *Pin notation and simplified block diagram of the LM1830 'fluid-level detector' IC*

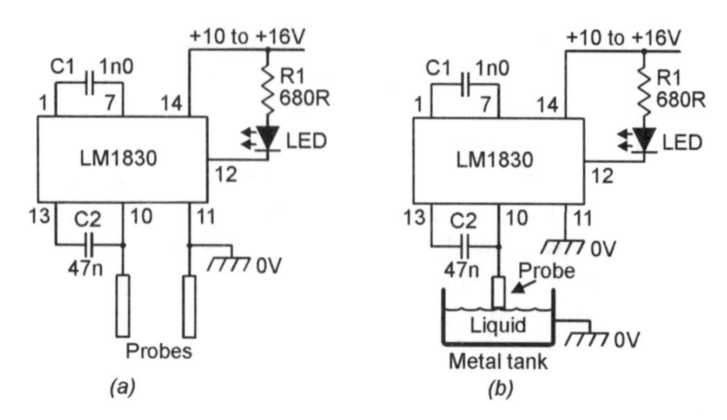

Figure 8.8 *Basic LM1830 low-fluid-level warning circuit with LED output, using (a) separate probes and (b) a single probe*

12) an open-collector npn common-emitter output stage that can sink up to 20mA maximum. The oscillator frequency is set via an external capacitor (1n0 gives 7kHz operation) wired between pins 1 and 7, and the IC can operate from supplies in the 9V to 25V range and consumes a typical standby current of 5.5mA.

Figure 8.8 shows the LM1830 IC's basic application circuits as a low-liquid-level alarm with an LED output; note that the *(a)* circuit uses two separate probes, one of which is grounded, but that in the *(b)* circuit – in which the liquid is stored in a metal tank – the metal storage tank is grounded and acts as one of the circuit's two 'probes'. The IC's internal oscillator is set at 7kHz via C1, and the non-grounded metal probe is taken to the pin-10 'detector' input and is ac driven via C2 and the internal 13k reference resistor. When the liquid level is 'high' (i.e. in contact with the probe) the probe-to-ground resistance is below the 13k reference value, and under this condition the output LED is off. When the liquid level is low the probe-to-ground resistance is high (greater than the 13k reference value), and under this condition the output LED is driven by a 7kHz squarewave signal and thus illuminates.

The basic *Figure 8.8* circuit can be usefully modified in a variety of ways, as shown in *Figures 8.9* to *8.12*. *Figure 8.9* shows it modified to give an audible 700Hz tone output (set by C1) into an inexpensive piezo sounder.

The LM1830 IC has a mean output current limit of 20mA maximum, and *Figure 8.10* shows how the available output power can be boosted via an external emitter-follower stage, and also shows how the IC can be used with an external (rather than the internal) reference resistor of up to 100k maximum (1k0 minimum), to enable it to test high-resistance liquids.

Figure 8.11 shows the circuit modified to give relay output drive via pnp emitter-follower Q1, by using C3 to convert the output-stage driving signal to

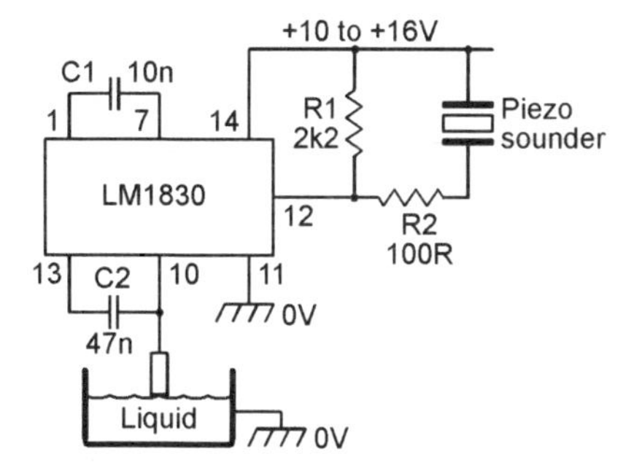

Figure 8.9 *Low-fluid-level warning circuit with 700Hz tone output*

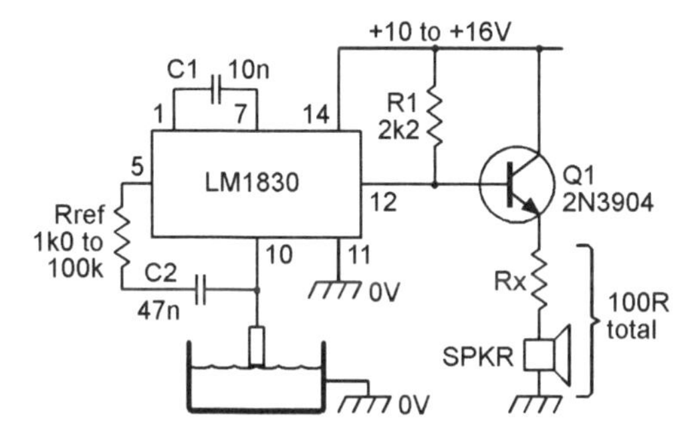

Figure 8.10 *Low-fluid-level warning circuit with external reference resistor and boosted audio output*

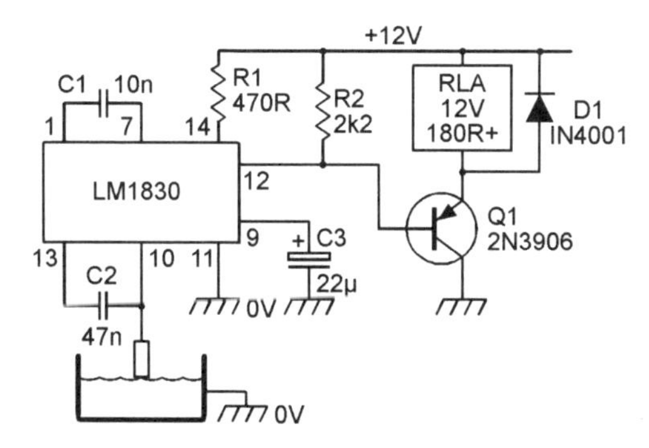

Figure 8.11 *Low-level warning circuit with relay output and supply-transient protection*

dc. This circuit also shows supply-line transient protection given to the IC via R1; this modification is recommended for use in automobile circuits, where – under very exceptional circumstances – supply transients may reach 40–50V.

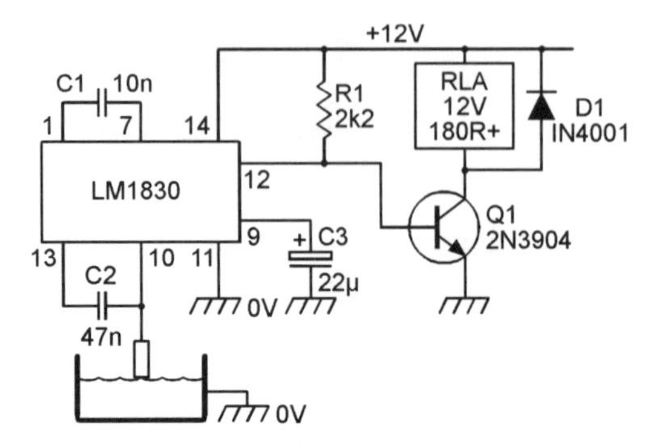

Figure 8.12 *Over-level warning circuit with relay output*

Finally, *Figure 8.12* shows the relay-driving circuit modified to give an over-level warning action (in which the relay is off when the liquid level is low) by using npn common-emitter amplifier Q1 as the relay driver.

'It's starting to rain' alarms

In households where washing is frequently hung up to dry in the open air, it is useful to have a device that gives an instant warning when the first few drops of rain start to fall as a prelude to a heavier shower. Rain can, for electronic detection purposes, be divided into two main types, which fall as either 'fine' or 'coarse' water globules. 'Fine' rain has a mist-like quality; its individual globules are relatively small and light, fall with relatively low velocity, and generate little audible noise. In the opening phase of a fine-rain shower, the globules fall with a fairly high drops-per-unit-area per-second density.

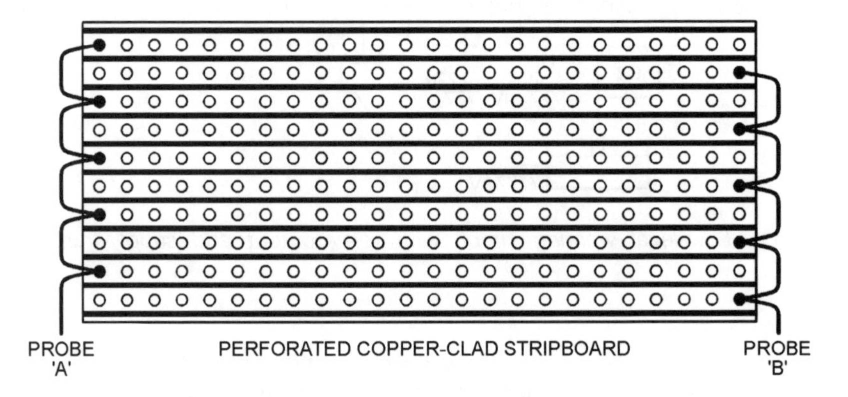

Figure 8.13 *Probe unit for use as a fine-rain sensor*

The easiest way to detect the onset of fine rain is to use a simple resistance-activated liquid-detector circuit (of one of the types already described) and use it in conjunction with a special probe unit made from perforated copper-clad stripboard in the manner shown in *Figure 8.13*. Here, alternate copper strips are wired together to make one probe, and the remaining strips are wired together to make the other probe, so that the upward-facing 'strip' probes are quickly shorted together (thus activating an alarm) by the first few drops of fine rain.

In practice, this strip-probe must be reasonably large, must be mounted clear of the ground (so that moisture is not trapped beneath it), must be kept clean and dry under no-rain conditions, must slope slightly downwards (so that rain runs off it) and should ideally be slightly heated, so that moisture evaporates from it within a few minutes.

Most rain is of the 'coarse' type, in which the individual globules are fairly large and heavy, fall with fairly high velocity, and generate considerable audible noise. In the opening phase of a coarse-rain shower, the globules fall with a very low drops-per-unit-area per-second density.

The easiest way to detect the onset of coarse rain is to use an acoustic 'sounding board' technique, in which the rain is allowed to fall on a large sheet of metal or hard plastic (or a similarly strong but flexible material) that, when hit by droplets of coarse rain, acts as a sounding board that magnifies the sound of each individual raindrop impact, thus producing a clearly audible sound below the sounding board. In the rain detector system, this sound is picked up by a microphone/transmitter unit, and the resulting audio signal is passed into the house via a 2-core cable and there reproduced by a high-gain audio power amplifier, thus giving a clearly audible warning of the rain's onset to any person that is listening. In practice, the sounding board can simply take the form of a large metal or hard-plastic box, lid, or container.

Figure 8.14 shows the practical circuit of a coarse-rain-detecting microphone/transmitter unit. Here, an electret microphone insert is used as an acoustic

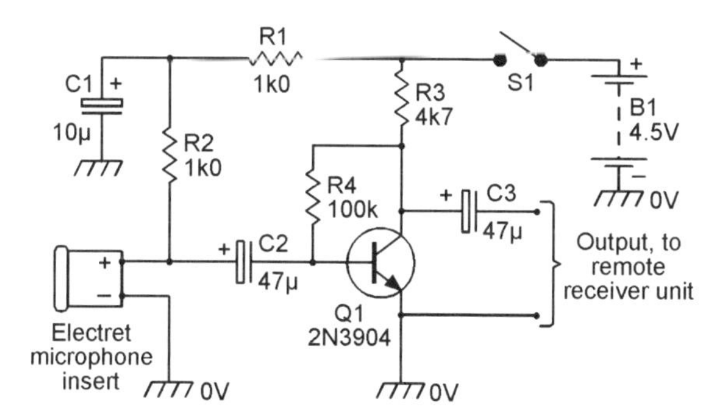

Figure 8.14 *Practical coarse-rain-detecting microphone/transmitter unit*

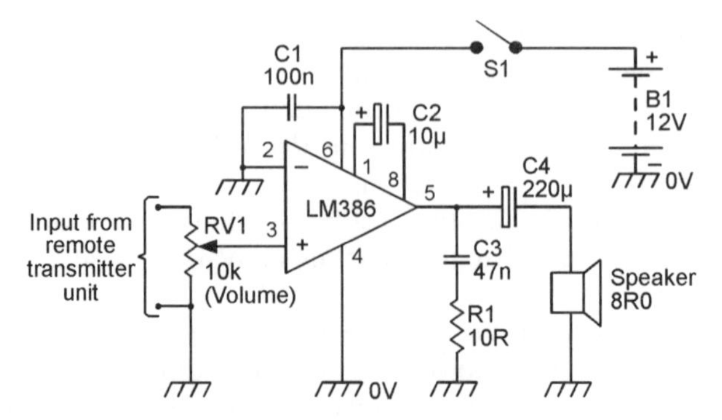

Figure 8.15 *'Coarse-rain-detector' audio power amplifier unit*

pick-up that uses R2 as its FET-drain load and has its audio output amplified by the Q1 common-emitter amplifier and passed on to the remote receiver unit via C3 and a 2-core cable. Note that the unit is powered via its own 4.5V battery, and that the microphone's supply line is decoupled via R1–C1. The complete unit consumes a typical operating current of about 1mA.

Figure 8.15 shows the circuit of the coarse-rain-detector's audio power amplifier or 'receiver' unit. Here, the input signals from the transmitter unit are applied across volume control RV1 and fed to the non-inverting input pin of a LM386 audio power amplifier IC, which has its voltage gain set at ×200 by C2 and (when powered from a 12V supply) can pump several hundred milliwatts of audio power into the 8R0 speaker load. Note that the unit is powered via its own 12V battery, and that C3–R1 form a stability-enhancing Zobel network.

A gas-activated alarm circuit

Leaking highly-flammable gases such as iso-butane, methane (natural or 'town' gas), hydrogen, and ethanol, etc., all present potentially explosive and life-threatening hazards, but can easily be detected – even in gas-to-air concentrations of less than 0.5% – by a simple and easy-to-use device that is readily available from major component suppliers and is known as a hot-wire gas sensor. The heart of this sensor is a coil of fine platinum wire that is coated with high temperature oxides and a special catalyser.

A hot-wire gas sensor actually consists of a *pair* of thermally matched hot-wire elements, one of which is gas-sensitive and is known as the 'detector', and the other of which is not gas-sensitive and is known as the 'compensator'. In some cases, the matched detector and compensator are supplied as individual units, which are each enclosed within an individual fire- and explosion-proof wire mesh, and in others they are combined in a single unit

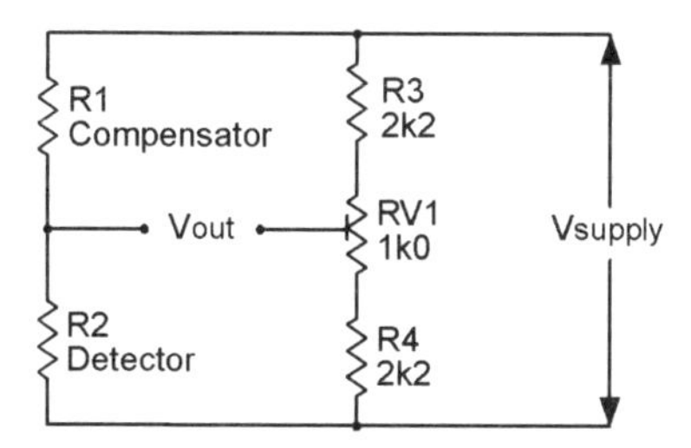

Figure 8.16 *Basic gas detector circuit using 'hot-wire' gas sensors*

and share a common fire/explosion-proof mesh. In either case, they are meant to be used in the basic circuit shown in *Figure 8.16*.

In *Figure 8.16* the compensator (R1) and detector (R2) are wired in series to form a gas-sensitive potential-divider on one side of a Wheatstone bridge, R3–RV1–R4 form an adjustable potential divider on the other side of the bridge, and RV1 is adjusted so that Vout is normally zero. The compensator and detector have a low hot-wire resistance, and when wired as shown and powered from a suitable voltage source (typically 2.2V or 3V AC or DC) pass a current (typically 150mA to 400mA) that raises the hot-wire temperature to about 350°C in gas-free air.

The resistances of the detector and compensator both vary with ambient temperature and humidity levels, etc., but are matched so that they vary equally in both devices, so that (when wired as shown in *Figure 8.16*) they maintain a constant division ratio in the absence of gas. When gas is present, the detector's special catalyser effectively but safely burns the gas that strays within the safety mesh and which surrounds the hot-wire, thus raising the hot-wire's temperature and resistance, thereby reducing the voltage appearing on the detector/compensator junction and upsetting the balance of the bridge. This action typically makes the circuit's Vout value fall by about 25mV at gas concentrations of 4000ppm (= 0.4%) with methane, or 2000ppm (= 0.2%) with isobutane.

Figure 8.17 shows the circuit of a practical gas alarm that is powered via an external 12V DC supply and which drives a ready-built solid-state commercial alarm/siren unit under the 'alarm' condition. Here, the basic gas detector (which is similar to that of the *Figure 8.16* circuit) is built around R2–RV2–R3–R4–R5 and is powered via a stable low-voltage DC supply derived from the 12V line via voltage regulator IC1, and has its output fed to the alarm/siren unit via voltage comparator IC2 and transistor switch Q1. The circuit's action is such that the voltage on pin-2 of IC2 is normally about 25mV above that of pin-3 (settable via RV2) under 'clean air' conditions, and under this condition IC2's output is low and Q1 and the alarm/siren are off, but the pin-2 voltage falls below that of pin-3 when significant gas concentrations are detected, and under this condition IC2's output switches high and drives Q1 and the alarm/siren on.

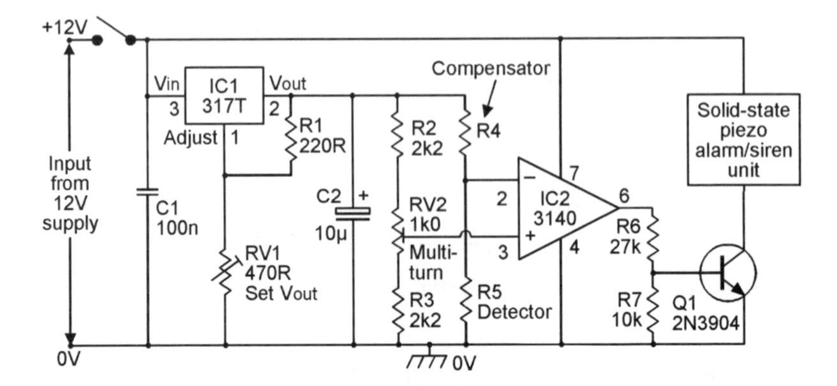

Figure 8.17 *Practical gas-detector alarm circuit, powered from an external 12V DC supply*

Before starting to build the *Figure 8.17* circuit, first locate a suitable hot-wire gas sensor and find out its voltage and current ratings. With that current rating in mind, select a suitable 12V DC power supply. Now – *without wiring the gas sensor in place* – build the IC1 voltage regulator section of the circuit, taking care to fit IC1 to a heat sink that will dissipate 1W per 100mA of working load current, and then power it up and trim RV1 so that (when powering a dummy load) it produces the precise specified working voltage of the gas sensor (usually 2.2V or 3V).

Now build the rest of the circuit, fit the gas sensor in place, power the circuit up, and trim RV2 so that the alarm/siren is off. Let the unit warm up for a minute or two, then – using a high-impedance digital multimeter – trim RV2 so that pin-2 of IC2 is 25mV above that of pin-3. The unit is now set and ready for use, and should activate the alarm if the sensor is temporarily placed in a box and exposed to a modest concentration of gas (such as a brief squirt of butane gas) for half a minute or so. Note that most flammable gases are heavier than air, and that in normal domestic situations the gas sensor should thus be mounted a few inches above floor level, in a position where it is unlikely to be damaged by passing feet or by the movement of furniture, etc.

A sound-activated switch circuit

Prior to the advent of modern highly reliable PIR movement detectors, sound-activated alarms were widely used in commercial security systems. Most of these alarms were, however, easily false-triggered by the natural sounds that occur inside buildings (such as the cracks or groans that occur as a building cools at night or warms up in the morning) or that originate

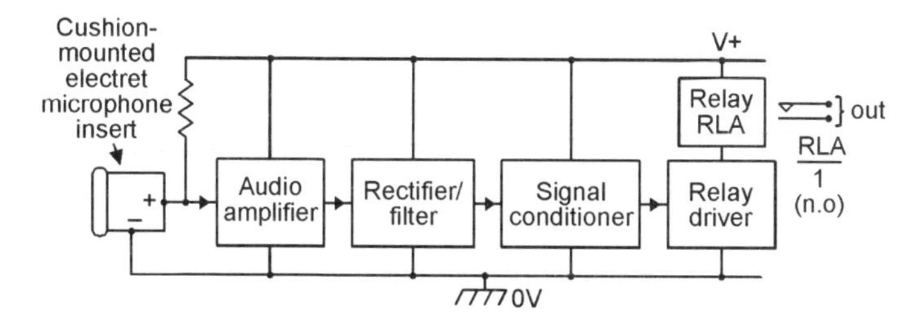

Figure 8.18 *Block diagram of a modern sound-activated switch circuit*

outside the buildings but are audible within them (such as loud traffic or aircraft noise or thunder). In some systems, the latter problem was overcome by using internal and external sound detectors and – by using sound-level comparison techniques – only activating the alarm if the internal sounds were louder than the external ones.

Today, sound-activated alarms are rarely used in security systems, but sound-activated *switches* are widely used. They are used mainly as relay-driving 'precautionary warning' devices that switch on a security light and/or a pre-recorded 'verbal warning' message or activate a sound- or video-recorder system whenever a suspicious sound is heard in a protected area. *Figure 8.18* shows the typical block diagram of a modern sound-activated switch circuit.

In *Figure 8.18*, sounds are picked up via a cushion-mounted electret microphone insert (which thus responds mainly to air-conducted – rather than structure-conducted – sounds) and are then amplified, converted to dc via a rectifier/filter, and then fed to a non-latching relay-driver via a special signal-conditioning circuit.

Figure 8.19 shows a practical relay-driving sound-activated switch circuit that is powered from a 12V DC supply. Here, the cushion-mounted electret

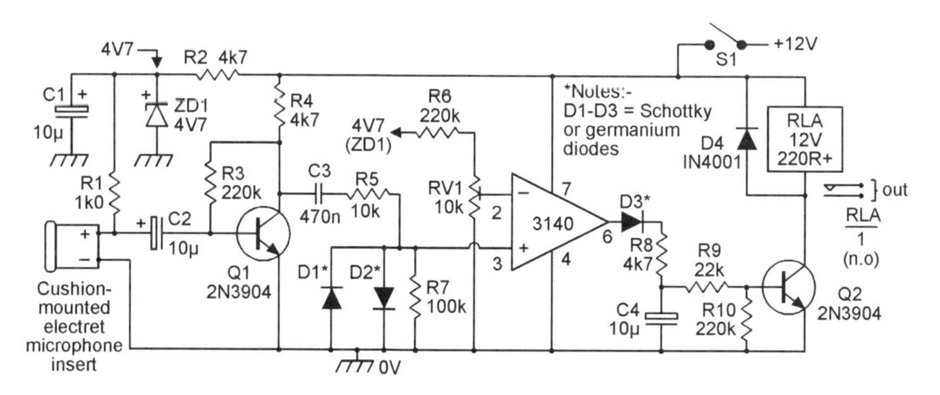

Figure 8.19 *Practical relay-driving sound-activated switch circuit*

microphone insert is powered from a stable 4V7 supply derived from the +12V line via Zener diode ZD1, and has its output amplified by common-emitter amplifier Q1 and then passed on to the pin-3 input of the 3140 op-amp via C3–R5. The 3140 op-amp can respond to input signals all the way down to zero volts, and in this circuit is used in the open-loop voltage comparator mode and acts as a super-efficient high-gain signal rectifier that has its 'threshold' level (and thus the circuit's sensitivity) fully variable from zero to +200mV via RV1; note that diodes D1–D2 act as clamps that limit the pin-3 peak-to-peak signal amplitudes to safe values, and *must* be germanium or Schottky (rather than silicon) signal diodes.

In *Figure 8.19*, D3–R8–C4–R9 and R10 act as the unit's special relay-driving signal-conditioning circuitry. Here, D3 and C4 peak-detect the pin-6 output voltage of the 3140 op-amp, and C4's resulting charge provides base drive to relay-driving common-emitter amplifier Q2. Note that C4's 'charge' time (which protects the circuit against activation by brief noise transients) is controlled by R8, and its 'discharge' time (which ensures that – once they have been triggered on – Q2 and RLA only turn off again when all noise trigger signals have been absent for a few seconds) is controlled by R9, and that these components provide the circuit with good immunity to false-triggering and relay-chatter problems.

To set up the *Figure 8.19* circuit, connect an analogue DC voltmeter between pins 6 and 4 of the 3140 op-amp, then trim RV1 so that the meter reading is zero at low sound input levels, but rises high enough to activate Q1 and RLA at the desired 'trigger' sound-amplitude level.

Power-failure alarm circuits

Electrical power-failure alarms can be made to activate when AC power is removed from a deep-freeze unit, or when a burglar deliberately cuts the AC power lines, or when a machine overloads and blows its fuses. Three useful power-failure alarm circuits are described in this section.

Figure 8.20 shows a simple relay-output power failure alarm that can be used to activate any type of external alarm device via the relay's contacts. Here, the power-line input is applied to a step-down transformer that gives an output of 12V at 100mA. This output is half-wave rectified by D1 and smoothed by C1, and the resulting DC directly powers the coil of relay RLA, which has a coil resistance of 220R or greater. RLA has one or more sets of n.c. change-over contacts that can be used to activate an external alarm device.

Thus, when AC power is applied to the *Figure 8.20* circuit the relay is driven on and contacts RLA/1 are open, and the alarm is thus off; the circuit typically consumes about 820mW from the AC power lines under this condition. When AC power is removed from the circuit, the relay turns off and its RLA/1 contacts close, thus activating the external alarm.

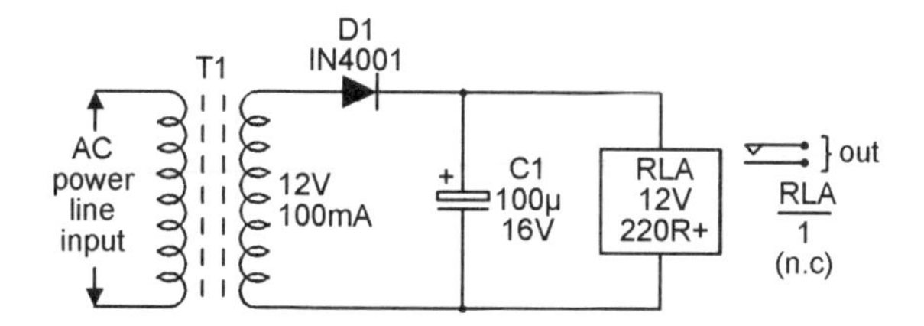

Figure 8.20 *Simple relay-output AC power-failure alarm*

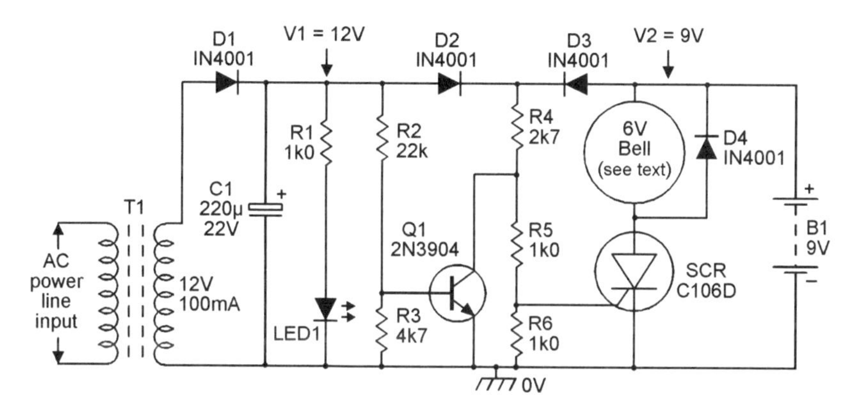

Figure 8.21 *Power-failure alarm with an alarm-bell output*

Figure 8.21 shows an alternative type of power-failure alarm. Here, the AC power input is stepped down to 12V by T1 and is rectified and smoothed by D1 and C1, to give roughly 12V DC at the D1–D2 and D2–D3 junctions. The actual alarm device, which is a self-interrupting 6V alarm bell that consumes less than 2A when operating, is used as the anode load of an SCR and is powered from a 9V battery.

When AC power is applied to the *Figure 8.21* circuit, roughly 12V DC is developed on the D1–D2 and D2–D3 junctions, and LED1 is illuminated via R1. Under this condition Q1 is driven to saturation via R2, and Q1's collector pulls the R4–R5 junction down to near-zero volts; zero drive is thus applied to the SCR gate, so the alarm is off and D3 is reverse-biased by the 12V on the D2–D3 junction and no current is drawn from the 9V battery. When AC power is removed from the circuit, R1–LED1 rapidly discharge C1, and the D1–D2 junction quickly falls to zero volts and Q1 turns off; under this condition current reaches the SCR gate from the 9V battery via D3–R4–R5, and the SCR and the alarm thus turn on.

Figure 8.22 shows a power-failure alarm that produces an output in a ready-built piezo siren unit. The circuit is similar to that of *Figure 8.21*, except that the SCR and alarm bell are replaced by a transistor and siren

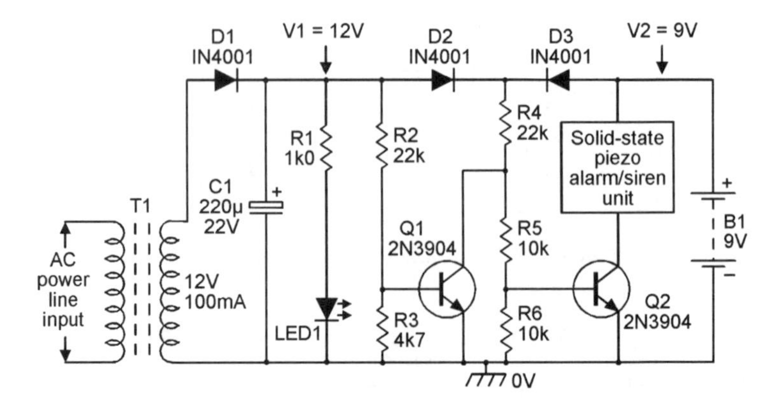

Figure 8.22 *Power-failure alarm with a piezo siren output*

unit. When AC power is applied to this circuit, LED1 illuminates and Q1 is driven to saturation via R2 and pulls the R4–R5 junction down to near-zero volts. Under this condition zero base drive is applied to Q2, so the piezo alarm is off and no current is drawn from the 9V battery. When AC power is removed from the circuit's input, R1–LED1 rapidly discharge C1, and the D1–D2 junction quickly falls to zero volts and Q1 turns off; under this condition current reaches Q2's base from the 9V battery via D3–R4–R5, and Q2 and the siren thus turns on.

Note that the *Figure 8.21* and *8.22* circuits can, if desired, be used with higher battery and T1-secondary voltages, provided that the resulting V1 voltage is at least 2V greater than V2 (the battery voltage).

A proximity-activated 'alarm' circuit

A proximity-activated alarm is a circuit that activates when a person or large object touches or comes close to a sensing antenna. The antenna may simply consist of a length or loop of wire, or may be a metal object (such as a sheet of foil or wire mesh hidden under a carpet, a safe, or a storage cabinet) that is connected to one end of a wire antenna.

Most proximity-activated alarm circuits work on the capacitive loading principle, in which the gain of an L–C oscillator is adjusted to a critical point at which oscillation is barely sustained, and in which the antenna forms part of the oscillator's tank circuit, and in which the circuit's 0V supply line is grounded. Consequently, any increase in the antenna-to-ground capacitance, such as is caused by touching or nearing the antenna, causes enough damping of the tank circuit to bring the oscillator gain below the critical level, and the oscillator ceases to operate. This cessation of oscillation is then used to make the alarm generator activate.

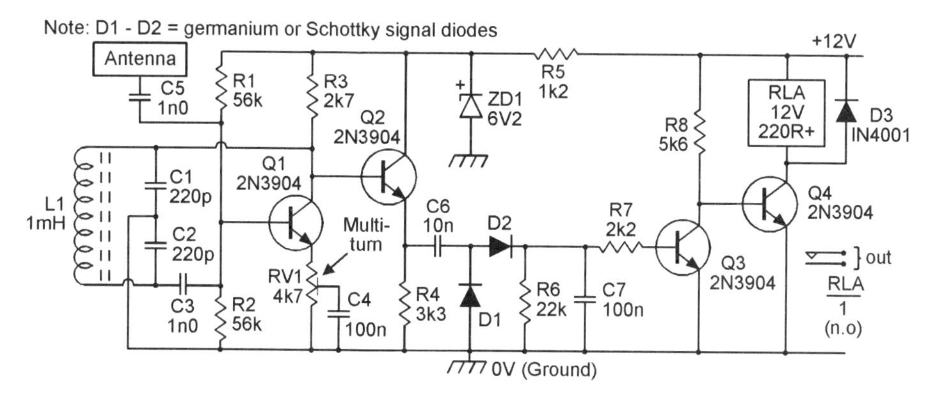

Figure 8.23 *Relay-output proximity-activated alarm circuit*

Figure 8.23 shows a practical relay-output proximity-activated alarm circuit that uses the above operating principle. Here, transistor Q1 is wired as a Colpitts oscillator, with gain adjustable via RV1, and the antenna is coupled to Q1 base via C5. The output of this oscillator, which operates at about 300kHz, is made available at a low impedance level across R4 via emitter-follower Q2. This signal is rectified and smoothed via the C1–D1–D2–R6–C7 network, to produce a positive bias that is applied to the base of Q3 via R7. Q3 is wired as a common-emitter amplifier, with R8 as its collector load, and Q4 is wired as a common-emitter amplifier with the relay coil used as its collector load and with Q4's base connected directly to the collector of Q3.

Thus, when the *Figure 8.23* circuit is operating normally, the oscillator output produces a positive bias voltage that drives Q3 to saturation and thus removes Q4's base drive; Q4 and the relay are thus off under this condition. When the circuit's antenna is touched or externally loaded, however, the oscillator ceases to operate, thus removing Q3's base drive; Q3 thus turns off; with Q3 off, Q4 is driven to saturation via R8, and the relay is thus driven on under this condition.

Note that the Q1–Q2 section of the circuit is powered via a 6.2V regulated supply formed by R5 and ZD1, thus enhancing the oscillator's stability. Also note that D1 and D2 must be germanium or Schottky signal diodes, and that the circuit can – if desired – be used to give a directly-driven 'siren' output (via a ready-built piezo alarm module) by simply using the siren module in place of RLA and removing D3 from the circuit.

To set up the *Figure 8.23* circuit, simply connect a suitable antenna, trim RV1 so that the relay turns on, then back RV1 off slightly so that the relay just turns off again. Check that the relay turns on again if the antenna is touched or closely approached, and goes off again if the touch is removed; if necessary, trim RV1 for maximum sensitivity.

The final sensitivity of the *Figure 8.23* circuit depends on the setting of RV1 and on the size and type of antenna used. If the antenna is very small, such as a short length of wire, the circuit will act as little more than a touch alarm, but if the antenna is a large sheet of metal foil or wire mesh, the circuit may be sensitive enough to activate when a person approaches within a foot or two of the antenna. It pays to experiment with different types of antenna, to get the 'feel' of the circuit. Remember, however, that the antenna must be well isolated from ground, and that the circuit's 0V rail must be wired to an effective ground connection.

Touch-activated circuits

Touch-activated circuits are intended to perform some kind of switching action when a person touches a fixed (rather than flexible or mobile) contact point, such as one or more metal studs. Circuits of this type may work in a variety of ways. Some work on the capacitive loading principle described in the preceding section of this chapter. Some are meant for use only in the general vicinity of AC power lines, and are activated by the power-line radiated AC 'hum' that is picked up by an electrical contact when it is touched by a human finger. Some are activated by the relatively low resistance (less than a few megohms) that appear across a pair of contacts when they are bridged by a human finger. Practical circuits of the latter two types are described in this section of the chapter.

Figure 8.24 shows a practical 'hum-detecting' touch-activated circuit that activates relay RLA when a finger touches a single metal stud or contact point; the circuit is designed around a CMOS 4001B quad 2-input NOR gate IC and one transistor. Here, gate IC1a is wired as a simple pulse-inverting amplifier and has its high-impedance input terminal taken to the metal touch contact via R2; the contact is biased high via R1, and IC1a's output is thus

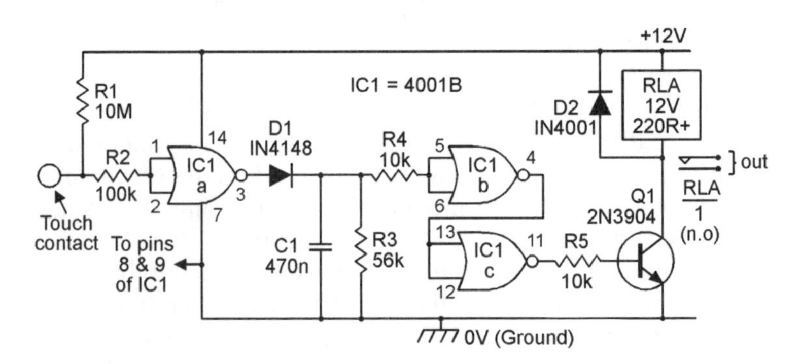

Figure 8.24 *'Hum-detecting' touch-activated relay switch*

normally low. When a human finger touches the circuit's contact terminal, its induced AC 'hum' signal reaches IC1a's input, is amplified and inverted, and appears as a large amplitude squarewave at IC1a's output. This square-wave is rectified and smoothed via D1–C1–R3, is buffered by gates IC1b–IC1c (which act together as a non-inverting buffer), and then drives relay RLA on via R5 and Q1.

Note when using the *Figure 8.24* circuit that its 12V supply must be derived (via an isolating transformer) from the AC power lines, that the 0V supply rail must be grounded, that the relay's contacts can be used to activate external circuitry or alarms, etc., and that the circuit consumes a quiescent current of only 1μA or so. The circuit's touch contact should not be larger than about ten square centimetres (to avoid unwanted pick-up); if the contact is more than a few inches from IC1a's input terminal, the connecting leads may have to be screened to avoid unwanted pick-up.

A simple resistance-sensing touch-activated relay switch circuit is shown in *Figure 8.25*. Here, a single 4001B gate is wired as a digital 'logic-level' inverter and has its input tied high by 10M resistor R1, so the gate's output (which is used to activate the relay via Q1) is normally low, and the relay is thus normally off. Note that IC1a's input is wired to one of a closely-spaced pair of contacts via R2, that the other contact is wired to the 0V rail, and that IC1's output goes high if a resistance lower than 10M appears across the contacts. Human skin normally has a resistance far lower than 10M. Consequently, if a human finger is pressed against both contacts simultaneously, IC1a's output switches high and turns RLA on via Q1.

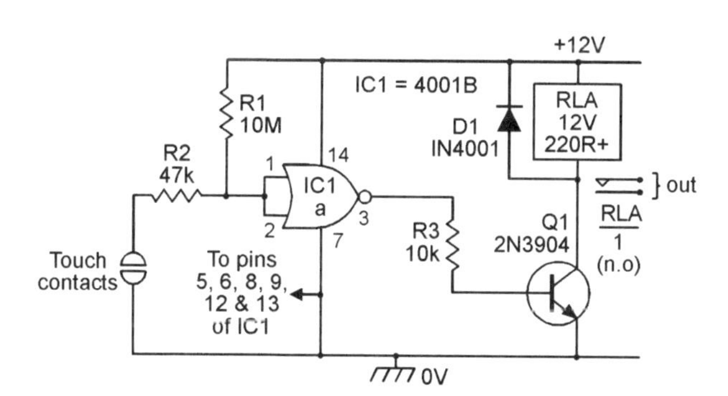

Figure 8.25 *Resistive touch-activated relay switch, normal contacts*

The *Figure 8.25* circuit consumes a typical quiescent current of less than 1μA. When building the circuit, note that the touch contacts should each have a surface area of at least half a square centimetre, and should be placed no more than half a centimetre apart.

Finally, *Figure 8.26* shows the above circuit modified for use with micro-sized contacts. The circuit is similar to that described above, except that R1

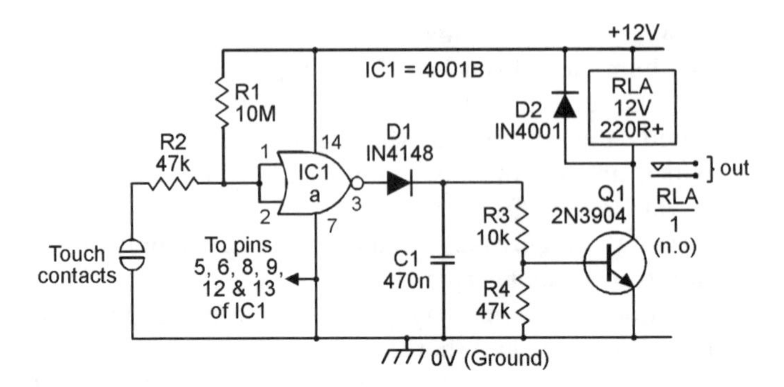

Figure 8.26 *Resistive touch-activated relay switch, micro-contacts*

is increased to 20M by wiring two 10M resistors in series, and that the design is also made sensitive to hum pick-up signals via D1 and C1. The sensitivity of this circuit is such that it can be used with pinhead-sized touch contacts.

An ultrasonic 'beam' alarm unit

This unit can be used in the same type of application as an IR light-beam alarm, but works on ultrasonic principles. It consists of an ultrasonic transmitter (Tx), operating at about 40kHz, which is aimed at a matching relay-driving ultrasonic receiver (Rx) unit; when an ultrasonic link exists between the Tx and Rx, the relay is off, but when the link is broken the relay turns on and activates an external alarm or some other electrical or electronic device. This particular unit is a very simple design, with a maximum operating range of only a few yards; it can be used to give security protection to passages and open doorways, etc.

The unit makes use of a modestly priced matched pair of ultrasonic transducers of the type used in many remote-control applications. These devices are normally designed to operate at about 40kHz, and consist of a dedicated Tx transducer and a matching dedicated Rx transducer; devices of this type are readily available from major electronic component suppliers.

Figure 8.27 shows the circuit of the unit's transmitter module, which typically consumes an operating current of 2.5mA from a 9V supply or 3mA from a 12V supply. Here, Q1 and Q2 are configured as an emitter-coupled oscillator, with the Tx transducer used as the emitter coupling element, so that the circuit oscillates at the transducer's resonant frequency (about 40kHz) and radiates a matching ultrasonic signal.

Figure 8.28 shows the circuit of the unit's receiver module. Here, the Rx transducer is pointed towards the transmitter and responds to the transmitted signal in much the same way as a directional microphone. The output of

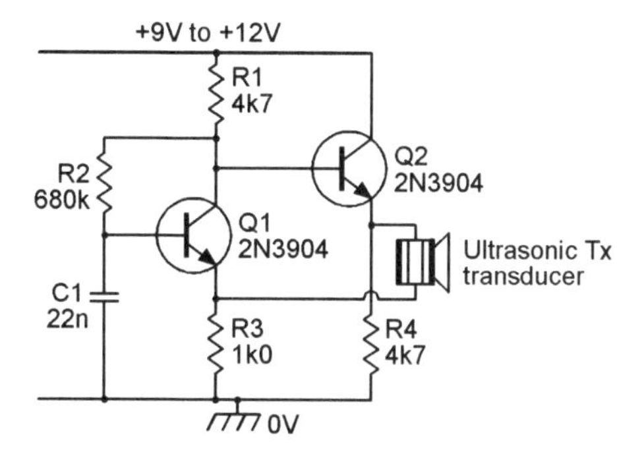

Figure 8.27 *Ultrasonic beam transmitter module*

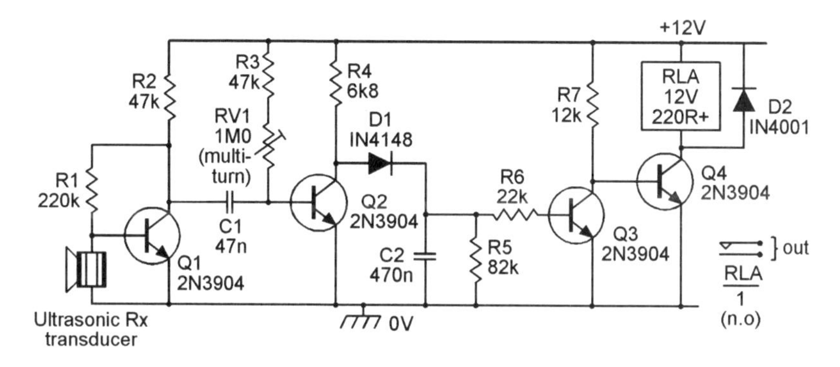

Figure 8.28 *Relay-output ultrasonic beam receiver module*

the Rx transducer is fed directly to the base of common emitter amplifier Q1, and appears in amplified form at Q1 collector. It is then fed, via C1, to the input of an amplifying detector stage that is built around Q2–D1 and C2. Normally, when the beam is unbroken, the output of this detector stage is high, so Q3 is driven to saturation and Q4 and the relay are cut off. When the beam is interrupted, the output of the detector stage falls to near-zero volts, so Q3 turns off and Q4 and the relay are turned on via R7. An external alarm can be activated by the closing of the RLA/1 contacts. Thus, RLA is normally off, but turns on when the ultrasonic beam is interrupted.

The *Figure 8.28* circuit consumes a typical current of 5mA from a 12V supply. To set up the circuit, turn off the Tx unit, connect a DC voltmeter (with a sensitivity of at least 20k/V) across C2, then trim RV1 so that the voltage just falls to near-zero; RLA should turn on under this condition. Now turn on the Tx unit, aim it at the Rx unit, and check that the C2 voltage rises to at least 2V and that RLA turn off. If desired, RV1 can be further trimmed to obtain absolute maximum operating range.

Eavesdropping circuits

Eavesdropping devices enable the user to secretly listen-in to private conversations or sounds. The most widely used type of electronic eavesdropping device is the miniature FM transmitter or 'bug', which incorporates a miniature electret microphone and a battery pack and usually has an effective transmission range of several hundred metres. In use, a standard or special-purpose FM receiver is tuned to the bug's operating frequency, and the bug is then secretly hidden in the room where the private conversations (etc.) are expected to take place. The user can then remotely listen-in to these private conversations via the FM receiver.

FM bugs are fairly readily available, either in kit or ready-built form, from various magazine advertisers in the UK and the USA and many other countries. It should be noted, however, that many of these bugs are poorly designed and break national wireless regulations, and that in some countries their use by ordinary citizens is illegal. It is now – as a result of current 'privacy' laws – quite illegal to use bugs *for the purpose of snooping* in the UK.

FM bugs come in a variety of basic types. They may be simple or sophisticated in design, may operate at a 'free' or a crystal-controlled frequency, may operate in the regular 88–108MHz FM band (where their Tx signals can be picked up on any standard FM receiver) or at some special VHF frequency (where their signals can only be picked up on a special FM receiver or on a scanner). True 'bugs' have a built-in microphone and audio amplifier and are designed to transmit voice signals, but 'pseudo-bugs' have the microphone and audio-amplifier replaced by an oscillator and are designed to transmit an alarm signal when activated. *Figure 8.29* shows a practical example of a simple pseudo-bug of the latter type, and *Figure 8.30* shows a very simple bug of the regular type.

The *Figure 8.29* circuit uses IC1 as a 1kHz squarewave generator that modulates the Q1 VHF oscillator and produces a harsh 1kHz tone signal in the FM receiver whenever the circuit's power supply is connected (such as by activating a hidden microswitch when a door opens, etc.).

The *Figure 8.30* circuit uses a 2-wire electret microphone insert to pick up voice sounds, etc., which are amplified by Q1 and used to modulate the Q2 VHF oscillator; this circuit thus acts as a regular FM bug. In both circuits the VHF oscillator is a Colpitts type but with the transistor used in the common-base mode, with C7 giving feedback from the tank output back to the emitter 'input'.

These two circuits have been designed to conform with American FCC regulations, and they thus produce a radiated field strength of less than 50μV/m at a range of 15 metres and can be freely used in the USA It should be noted, however, that their use is illegal in most other countries, including the UK.

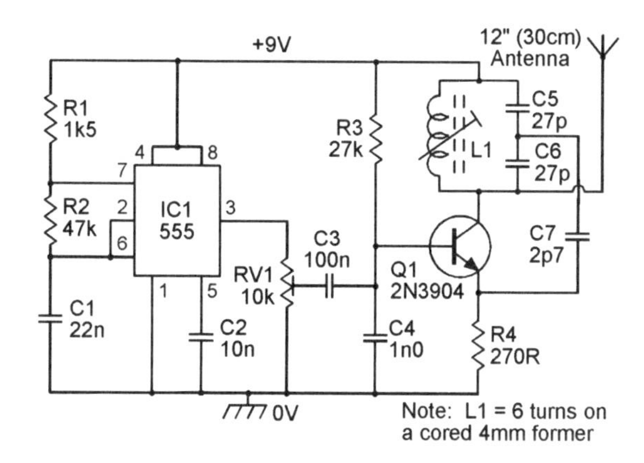

Figure 8.29 *FM 'pseudo-bug' transmits an alarm signal*

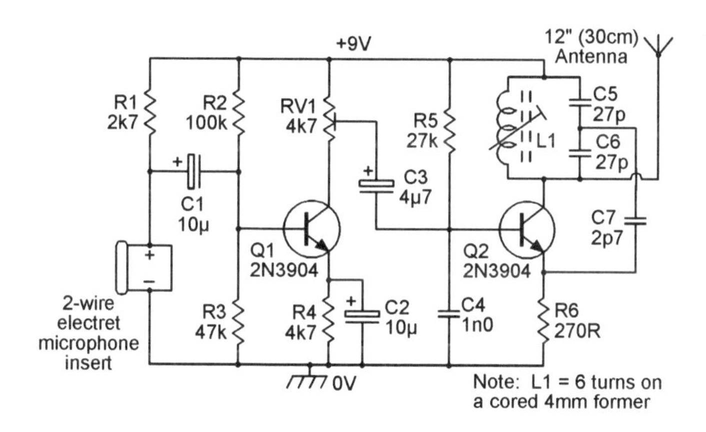

Figure 8.30 *Simple FM 'bug' transmits microphone signals*

To set up the *Figure 8.29* and *8.30* circuits, set the coil slug at its middle position, connect the battery, and tune the FM receiver to locate the transmitter frequency. If necessary, trim the slug to tune the transmitter to a clear spot in the FM band. RV1 should then be trimmed to set the modulation at a 'clean' level.

Basic CCTV applications

Closed circuit television (CCTV) systems are among the most useful of all modern security devices. In their simplest form, they consist of a fairly inexpensive black-and-white TV camera that is mounted in a fixed position and can be monitored on an ordinary domestic TV set. At the other end of the scale, the system may use an array of high-definition colour or infra-red TV cameras, each fitted with remote-controlled zooming and panning facilities, with all of the camera outputs fed to banks of TV monitors and video recorders.

CCTV systems have excellent crime-deterrent value, and in most crime-prevention applications each camera should be prominently displayed and fitted with an attention-grabbing flashing LED (cheap dummy cameras are also manufactured for use as crime deterrents). In domestic and small-business applications, one or several cameras can be hooked up to a normal TV set via a multi-function selector box that allows the user to watch normal TV programmes or to monitor individually-selectable camera outputs, or to automatically cycle through all available camera outputs or simultaneously display several camera outputs, at will. Some modern cameras are fitted with a PIR movement detector that automatically gives that camera priority position in the monitoring system if any kind of intrusion is detected.

In many commercial applications, several cameras may be in continuous use during business hours, and are used both as a deterrent and as a means of recording any crime that occurs (so that the criminals may later be identified via a video recording). In such systems, the output of each camera may be continuously fed to a special video recorder that uses standard recording tapes but uses high quality automatic freeze-frame (rather than continuous) recording techniques, thus enabling eight or more hours of useful recording to be stored on a single tape.

In some special applications (such as when trying to catch a serial sneak thief) it is useful to be able to monitor an area with a covert TV camera, and a variety of small 'pin hole' TV cameras are available for this purpose. Some of these cameras come disguised as an every-day item such as a clock, a smoke alarm, or a simple PIR unit, etc.

Index

Integrated circuits by type number